BASIC
STEEL DESIGN

Twin Towers of the World Trade Center during construction.
(Courtesy Port Authority of New York and New Jersey.)

BASIC
STEEL DESIGN

BRUCE G. JOHNSTON

Professor Emeritus, University of Michigan
Lecturer, University of Arizona

FUNG-JEN LIN

Senior Structural Engineer, The Ralph M. Parsons Company, Pasadena

Prentice-Hall, Inc.

Englewood Cliffs, New Jersey

Library of Congress Cataloging in Publication Data

JOHNSTON, BRUCE GILBERT.
   Basic steel design.

    1. Building, Iron and steel.   2. Structural design.
I. Lin, Fung-Jen, joint author. II. Title.
TA684.J6   624'.1821   73-1906
ISBN 0-13-069351-0

Civil Engineering and Engineering Mechanics Series

N.M. Newmark and W. J. Hall, Editors

10 9 8 7 6 5 4 3 2

Printed in the United States of America

PRENTICE-HALL INTERNATIONAL, INC., *London*
PRENTICE-HALL OF AUSTRALIA, PTY. LTD., *Sydney*
PRENTICE-HALL OF CANADA, LTD., *Toronto*
PRENTICE-HALL OF INDIA PRIVATE LIMITED, *New Delhi*
PRENTICE-HALL OF JAPAN, INC., *Tokyo*

CONTENTS

Preface   viii

Chapter 1 **The Steel Structure**   1

   1.1   Introduction   1
   1.2   The Structure and its Parts   2
   1.3   Structural Steel   4
   1.4   Structural Design Developments   6
   1.5   Structural Design Economy   9
   1.6   Structural Safety   10
   1.7   Planning and Site Exploration for the Specific Structure 11
   1.8   Structural Layout, Details, and Drawings   12
   1.9   Fabrication Methods   13
  1.10   Construction Methods   14
  1.11   Service and Maintenance Requirements   15

Chapter 2 **Tension Members**   17

   2.1   Introduction   17
   2.2   Types of Tension Members   19
   2.3   Allowable Tensile Stress   25
   2.4   Design for Repeated Load   26
   2.5   Flow Chart   27
   2.6   Illustrative Examples   28

Chapter 3 **Beams**   36

    3.1 Introduction   36
    3.2 Elastic Bending of Steel Beams   41
    3.3 Inelastic Behavior of Steel Beams   44
    3.4 Allowable Stresses for Elastic Design   47
    3.5 Lateral Support Requirements   54
    3.6 Beam Deflection Limitations   55
    3.7 Beams Under Repeated Load   56
    3.8 Biaxial Bending of Beams   56
    3.9 Load and Support Details   57
  3.10 Allowable Load Tables for Beams   59
  3.11 Flow Charts for Steel Beam Design   59
  3.12 Illustrative Examples   65

Chapter 4 **Columns under Axial Load**   82

    4.1 Introduction   82
    4.2 Basic Column Strength   83
    4.3 Effective Length of Columns   86
    4.4 Types of Steel Columns   88
    4.5 Width–Thickness Ratios   94
    4.6 Column Base Plates and Splices   95
    4.7 Allowable Compression Stress   95
    4.8 Flow Charts   97
    4.9 Illustrative Examples   99

Chapter 5 **Columns under Combined Stress**   111

    5.1 Introduction   111
    5.2 Allowable-Stress Design   112
    5.3 Design by Use of Interaction Formulas   113
    5.4 Equivalent Axial Compression Load   115
    5.5 Flow Charts   117
    5.6 Illustrative Examples   121

Chapter 6 **Connections**   128

    6.1 Introduction   128
    6.2 Riveted and Bolted Connections   129
    6.3 Pinned Connections   138
    6.4 Welded Connections   140

6.5   Eccentrically Loaded Connections   150
6.6   Shear Connections for Building Frames   157
6.7   Moment-Resisting Connections   167
6.8   Concluding Remarks Concerning Connections   174

Chapter 7   **Plate Girders**   178

7.1   Introduction   178
7.2   Selection of Girder Web Plate   180
7.3   Selection of Girder Flanges   183
7.4   Intermediate Stiffeners   189
7.5   Bearing Stiffeners   193
7.6   Connections of Girder Elements   195
7.7   Illustrative Example   197

Chapter 8   **Special Topics in Beam Design**   209

8.1   Introduction   209
8.2   Torsion   209
8.3   Combined Bending and Torsion   213
8.4   Biaxial Bending and Lateral-Torsional Buckling   223
8.5   Shear Center   232
8.6   Composite Design   235
8.7   Continuous Beams   235

Index   244

PREFACE

This book is written for the beginner in structural steel design. It is suitable for reference use or as a text. It is unique in at least two respects: (1) Reference is made primarily to a single design specification—that of the American Institute of Steel Construction, and (2) The chapters on individual member design include flow diagrams, similar to those used to guide the preparation of a computer program. These diagrams have been found to be excellent teaching aids.

The use of theory and stress analysis is minimized. The purpose of design is to produce a structure. In the design of complex and monumental structures, elaborate analyses, in many cases requiring use of the computer, may be required. But in the initial study of steel design, the acquisition of a basic understanding of structural behavior and the meaning of specification requirements can best be attained by simplicity of approach and emphasis on the development of sound structural judgment.

Chapter 1 is a broad and descriptive introduction to the steel structure, covering the properties of steel, the history of the development of steel structures, and touching on the topics of economy, safety, planning, fabrication, construction, and maintenance.

Chapters 2 through 7 are devoted to the various types of structural members in common use: the tension member, the beam, the column, and so on. Each of these chapters takes up the structural behavior problem, explains pertinent AISC Specification clauses, and summarizes (with the exception of Chapter 6) the logical application of the specification by means of a flow diagram. Although the flow diagram was developed primarily as an aid to the preparation of a computer program, it also serves admirably as a summary and as a guide to the logical sequence of steps that must be taken in the design selection of a particular structural member.

Chapter 8 covers special topics that are of occasional importance, but less common than those treated in the earlier chapters. These topics include the torsion of both open and box members, combined bending and torsion, biaxial bending, lateral-torsional buckling, the shear center, and a brief treatment on continuous beams with an introduction to plastic design. The material in Chapter 8 should be included in a first course if time permits.

Steel design specifications are essentially similar, but the diversity of formulas pertaining to identical problems (such as the column design formulas) is confusing to the beginning student. But after learning to design with a particular specification the student can readily adapt his basic knowledge to a different one.

The purchase of the Steel Construction Manual of the American Institute of Steel Construction is essential to the complete use and understanding of this book. The AISC Manual also includes the AISC Specification which will be referred to throughout the book. Moreover, practice in use of the AISC Manual is essential as a secondary educational objective toward structural design practice. The nomenclature that will be used herein is nearly identical with that found in the AISC Specification, pages 5–8 to 5–10 inclusive, and will not be repeated herein, except in individual references to the presentation of equations or formulas.

The flow chart symbols used in Chapters 2 through 7 have the following significance:

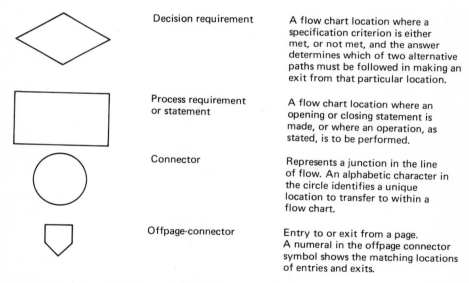

Symbol	Name	Description
(diamond)	Decision requirement	A flow chart location where a specification criterion is either met, or not met, and the answer determines which of two alternative paths must be followed in making an exit from that particular location.
(rectangle)	Process requirement or statement	A flow chart location where an opening or closing statement is made, or where an operation, as stated, is to be performed.
(circle)	Connector	Represents a junction in the line of flow. An alphabetic character in the circle identifies a unique location to transfer to within a flow chart.
(offpage symbol)	Offpage-connector	Entry to or exit from a page. A numeral in the offpage connector symbol shows the matching locations of entries and exits.

Special acknowledgement is made to William Milek, Jr., of the American Institute of Steel Construction, for his cooperation in the preparation of this book, and to Frank W. Stockwell, Jr., also of AISC, for his detailed check of much of the final draft. Thanks are also due to Randolph F. Thomas for his help in the preparation of many of the drawings. The authors have appreciated the support of their wives, Ruth and Hopi, and the first author will always be indebted to his father, the late Sterling Johnston, for having inculcated from an early age an empathy with steel structures. The Prentice-Hall staff, through Miss Nancy Baker, has been most cooperative in expediting the processing of the manuscript.

Each chapter includes a number of example problems which are presented with more complete details than would be required by an experienced designer. Problems for assignment are also included, with initial emphasis on variations of the example problems—thus adding incentive to the careful study of the examples in the text.

BRUCE G. JOHNSTON
FUNG-JEN LIN

1

THE STEEL STRUCTURE

1.1 INTRODUCTION

It is only by means of *structure* that the observable external details of our planet's surface are altered. Structures are the earmarks of our civilization, and the structural engineer—through the practice of construction within the framework of civil engineering—helps to create them: the buildings, dams, bridges, power plants, and towers that make possible our shelter, power, transportation, and communication. Thus the civil engineer has a responsible role in determining whether or not the structures that he builds enhance or detract from their environment.

After the prospective owner of a structure has considered alternatives and selected the site and has made subsurface exploration of soil conditions, the structural design is initiated by a consideration of various structural systems, alternative types and disposition of members, and the preparation of preliminary design drawings. Subsequently, the structural designer determines the required sizes of members and their connections, describing these in detail through drawings and written notes, so as to facilitate the fabrication and construction of the structural frame. One must first learn to design the parts before he can plan the whole. Hence the emphasis herein is on the design and selection of steel tension members, beams, compression members (columns), beam-columns, plate girders, and connections that join these members to form a bridge, building, tower, or other steel structure.

The adequacy of a structural member is in part determined by a set of design rules, called specifications, which include formulas that guide the designer in checking the strength, stiffness, proportions, and other criteria that may govern the acceptability of the member. There are a variety of specifications that have been developed for both materials and structures. Each is based on years of prior experience gained through actual structural usage. The diversity of specification formulas and rules pertaining to essentially similar problems is a source of confusion to the beginning student of structural design. In this book reference will be made primarily to a single specification—the widely used American Institute of Steel Construction (AISC) *Specifications for the Design, Fabrication and Erection of Structural Steel for Buildings*, as adopted in February, 1969, and supplemented in 1970 and 1971 by Supplements No. 1 and No. 2. He who masters the use of this specification, and understands the structural meaning and significance of its requirements, can readily turn to some other specification pertaining to the design of steel structures and quickly grasp the parallel set of design rules that it will contain.

The 1969 AISC Specification will be found in the seventh edition of the AISC *Manual of Steel Construction* along with much additional design information and tabular data. This AISC Manual must be considered to be an essential companion volume to this book and frequent reference will be made to it. To abbreviate the repeated references that will be made to the manual and the specification, they will be referred to herein as AISCM and AISCS, respectively.

At this point one should read the Foreword and Preface in the AISCM and thumb briefly through the entire book to get a preliminary idea of its contents.

1.2   THE STRUCTURE AND ITS PARTS

The basic framework that gives strength and form to a structure does so in the same way that the human skeleton gives strength and form to the human body, and in accordance with the same principles.

The creation of the complete structure calls for the combined services of the architect, civil engineer, and other specialists in engineering fields that may include acoustics, machine design, illumination, heat, ventilation, and other facilities. The overall design, processing, and scheduling of these inputs and the consideration of the involved interrelations while a structure is being planned and built have become known as *systems engineering*.

A primer for steel design must be concerned primarily with structural members that are the component parts of the overall structure. In a steel structure these are beams, which carry loads transverse to their long axis, columns or compression members, which transmit compression force along their long axis (the trunk of a tree is a most efficient column), and tension members, exemplified by a wire rope, most effectively capable of transmitting tension force, or pull, and built up of many individual wires that have been cold drawn to greatly increase their strength. Compression members usually must also carry some transverse loads and as such are called *beam-columns*. The way a structure is made up of these component parts is illustrated in Fig. 1.1 in which the upper part of a building frame is carried over an auditorium by means of a truss. In this figure the columns, beams, beam-columns, and tension

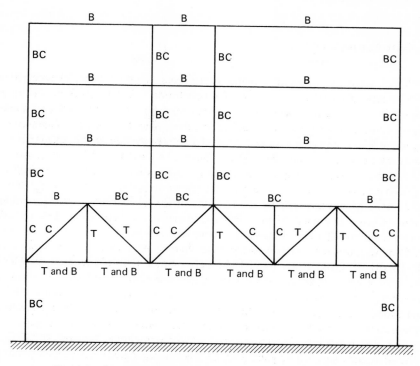

Fig. 1.1   Structural frames are composed of beams (B), columns
(C), tension members (T), and beam-columns (BC).

members are labeled by the letters C, B, BC, and T, respectively. At every juncture
point or joint between the ends of the members, connections must be provided, often
offering the most challenging design problems, as they are the least standardized—yet
essential to the continuity of the structure and its resistance to collapse.

A good structural designer *thinks* about the actual structure as much or more than
the mathematical model that he uses to check the internal forces for which he will
determine the required material and type, size, and location of the members that
carry the loads. The "structural engineering mind" is one that can visualize the real
structure, the loads upon it, and in a sense "feel" how these loads are transmitted
through the various members down into the foundations. Great designers are gifted
with what has sometimes been called "structural intuition." To develop "intuition"
and "feel," the engineer is a keen observer of other structures. He may even contem-
plate the behavior of a tree, designed by nature to withstand violent storms, flexible
where it is frail in leaves and small branches, but growing in strength and never aban-
doning continuity as the branches merge with the trunk, which in turn spreads below
its base into the root system that provides its foundation and connection with
the earth.

1.3 STRUCTURAL STEEL†

A knowledge of the elastic, inelastic, fracture, and fatigue characteristics of a metal is needed for the evaluation of its suitability for the making of a structural member for a particular structural application. *Elasticity* is the ability of a metal to return to its original shape after loading and subsequent unloading. *Fatigue* of a metal occurs when it is repeatedly stressed above its *endurance limit* through many cycles of loading and unloading. *Ductility* is the ability to be deformed without fracture in the inelastic range—that is, beyond the *elastic limit*. In steel, loaded in a simple tension state of stress, there occurs a sharp *yield point* at a stress only slightly greater than the elastic limit. Minimum specification values of the yield point, ductility indices, and chemistry have been established by the American Society for Testing Materials (ASTM) to control the acceptance of structural steels.

The mechanical properties of structural steel that describe its strength, ductility, and so forth, are given in terms of the behavior in a simple tension test. The initial portion of a typical tension stress–strain curve for structural steel is shown in Fig. 1.2(b). To a greatly different horizontal scale, the complete curve is given in Fig. 1.2(a), but the load-carrying capacity of beams and columns is very largely determined within the range of Fig. 1.2(b). The slope of the stress–strain curve in the elastic range is termed E, the modulus of elasticity, and is taken as 29,000 kips per square inch (ksi) for the structural steels. The yield point, F_y, is the most significant property that differentiates the structural steels for which the AISCS is applicable. Many of the specification requirements are written as formulas in which the term F_y appears, and Appendix A of the specification tabulates these requirements numerically for nine currently used yield points: 36, 42, 45, 50, 55, 60, 65, 90, and 100 ksi. The ultimate or breaking strength in tension, F_u, based on the original cross-sectional area, is also recorded for the tension test.

The yield point of steel will vary somewhat with temperature, speed of test, and the characteristics (size, shape, and surface finish) of the test specimen. After initial yield, the specimen elongates in the *plastic* range without appreciable change in stress. Actually, yield occurs at very localized regions, which *strain-harden*, that is, strengthen, so as to force yielding into a new location. After all the elastic regions have been exhausted, at strains of from 4 to 10 times the elastic strain, the stress starts to increase and a more general strain hardening or strengthening commences. The sharp yield point and flat yield stress level shown in Fig. 1.2(a) are peculiar to the non-heat treated structural steels.

Structural steels are unique in that they are tough. *Toughness* may be defined as a combination of strength and ductility. After the general strain-hardening range commences in the tension test, the stress continues to increase, and inelastic extension of

†The reader may wish to supplement the study of this section by reference to Chapter 1, "The Structural Steels and Their Mechanical Properties," of Reference 1.7.

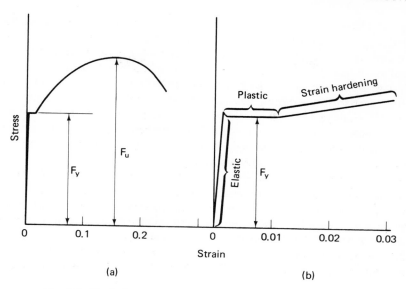

Fig. 1.2 Typical stress-strain curves from a tension test of structural steel.

the test specimen continues uniformly (without local reduction in cross-sectional area) until the maximum load is reached. The specimen then experiences a local constriction and is said to *neck down*. The nominal stress based on the original area is termed the *ultimate tensile strength* of the material. The ability of steel to withstand inelastic deformation without fracture also permits it to sustain local yielding during fabrication and construction, thus allowing it to be sheared, punched, bent, and hammered without apparent damage.

Under certain combinations of circumstances steel structures may develop cracks without appreciable prior ductile deformation. The designer should avoid sharp reentry corners that cause stress concentrations, especially in large boxlike or tank structures. Sheared edges and punched holes also cause minute stress concentrations and damage the edge material where cracks are apt to start. Operation in extremely low temperature is another factor conducive to brittle fracture. Thus careful attention to smooth edge preparation, avoidance of stress raisers, and quality control of material and fabrication processes will reduce the likelihood of brittle fracture to a minimum.

Most frequently the designer will use a standard steel shape as a structural member. These are hot rolled from billets, and their standardized dimensions for detailing and properties for design are completely tabulated in Part 1 of the AISCM. These include wide-flange and miscellaneous beams, channels, and angles. The beams range from a depth of 3 inches (in.) up to 36 in. at 300 pounds per foot (lb/ft), and include (primarily for use in tall buildings) a series of very wide column sections having a nominal depth of 14 in. with weights of 87 to 730 lb/ft.

The reader should turn to the AISCM pages 1-3 to 1-4, where the availability and selection of the appropriate grade of steel are discussed. On pages 1-4 to 1-9 the availability of shapes, plates, and bars are discussed and tabulated. The abbreviations and symbols used in designating hot-rolled structural steel shapes are listed in AISCM, page 1-10. They will be used throughout the text and in assigned problems.

In some parts of the country, less accessible than others to those mills that roll heavy shapes, equivalent shapes are made up by welding three plates together. Of course, when either beam or column section requirements exceed those available in standard rolled shapes, such sections are tailor made, as it were, by welding together plates to form heavy girder or column sections.

In addition to hot-rolled shapes, standard sizes of plate, bar, pipe, and hot-formed tubing, either square, rectangular, or circular in cross section, are available.

Supplementing the available range of hot-rolled sections is a wide variety of both standard and special cold-formed shapes. Their use in design is covered by the *Specification for the Design of Cold-Formed Steel Structural Members* [1.4] of the American Iron and Steel Institute, which is coordinated as much as possible with the AISCS.

Quoting from the commentary [1.5] on the cold-formed member specification,

> . . . members are cold-formed, in rolls or brakes, from flat steel, generally not thicker than $\frac{1}{2}$ and as thin as about 0.0149 in.
>
> Cold-formed members, as distinct from heavier, hot-rolled sections, are used essentially in three situations: (1) where moderate loads and spans render the thicker, hot-rolled shapes uneconomical, (2) where, regardless of thickness, members are wanted of cross-sectional configurations which cannot economically be produced by hot-rolling or by welding of flat plates, and (3) where it is desired that load-carrying members also provide useful surfaces, such as in floor and wall panels, roof decks, and the like.

1.4 STRUCTURAL DESIGN DEVELOPMENTS

The essential task of the structural designer is simply that of determining the member shape, size, and arrangement that will safely carry the loads which the structure is expected to experience. These loads may consist only of the dead weight of the structure itself or there may be added loads due to stored material, people, vehicles, snow, ice, wind, explosive blast, water currents, impact, and shock of earthquake—all dependent on the type of structure, its intended use, and its geographic location. Loads can occur in combination, and the probability of such combinations as well as the magnitude of the loads must be considered. If the location involves both potential hurricanes and earthquake, each of which may occur respectively for only hours or seconds once or twice in a lifetime, it would be false economy to design all structures for a full combination of the two. An exception might be made if the structure housed a key communication center on which the safety of many people depended.

Structural design in ancient times was simply a matter of repeating what had been done in the past, without knowledge of material behavior or structural theory, with success or failure determined simply on the basis of whether the building or bridge supported the actual load or collapsed under it. Experience then was the only teacher; it is still today a most important element of good design. Gradually, through centuries of experience, the art of proportioning members evolved. Empirical rules were established. The columns in Grecian temples were said to be proportioned with the slenderness ratio of a woman's leg. The great builders of the Renaissance had no knowledge of stress analyses, yet achieved structures that required more than empiricism. They were artist, architect, engineer, and builder combined, and their cathedral domes stand now as evidence that they were able to intuitively design structures that today would not be attempted without first solving an eighth-order differential equation.

Structures of the past and present, and predictions regarding structures of the future, are directly conditioned by the development and commercial availability of structural engineering materials. Certain of these materials, such as stone, brick, timber, and rope, have been used since the beginning of recorded history. Columns of stone blocks, hewn with precision, are dominant features of Egyptian, Greek, and Roman temples. The aquedects and bridges of Rome were stone arches, which, like columns, transmit primarily compressive stresses. The Stone Age of structures persisted into the early part of the nineteenth century, when most arches and domes were still built of stone masonry and held in place by stone buttresses.

The commercial development of iron provided the first of the structural metals that were to open up an entirely new world to the structural engineer. The first bridge to be constructed completely of cast iron in 1779 still stands at Coalbrookdale in England. But the use of cast iron, which failed with a brittle fracture in tension, was short-lived. The commercial production of wrought-iron shapes in 1783 brought rapid changes, as it made available a product with that added quality of toughness exemplified by an ability to take large tensile deformation in the inelastic range without fracture. Moreover, wrought iron could be formed into flat plates that could be bent and joined by rivets, making possible the steam locomotive, which, in turn, created a demand for long-span metal bridges. Noteworthy among the early wrought-iron bridges was the Britannia Bridge across the Menai Straits of the Irish Sea. It consists of twin parallel box girders continuous over four spans, with two center spans of 460 ft each, flanked by 230-ft end spans. It was completed in 1850 and is the prototype of a current trend in bridge construction that may be called "the rebirth of the box girder bridge."

The development of the Bessemer converter in 1856 and the open-hearth furnace in 1867 introduced structural steel, and this is the material that has been used most in bridges, as well as in many buildings, for the past 100 years. The first major bridge to be constructed entirely of structural steel was the famous Eads Bridge across the Mississippi at St. Louis. Completed in 1874, it incorporates tubular steel arches with a central span of 520 ft, flanked by 502-ft side spans.

Paralleling the development of iron and steel as engineering materials were advances in material-testing techniques and in structural analysis that made possible the transition of structural design from an art to an applied science. Hooke (1660) developed the concept that load and deformation were proportional, and Bernoulli (1705) introduced the concept that the resistance of a beam in bending is proportional to the curvature of the beam. Bernoulli passed this concept on to Euler, who in 1744 determined the elastic curve of a slender column under compressive load. Important developments in the late 1800s included: (1) Manufacture of mechanical strain-measuring instruments that made possible the determination of the elastic moduli that related stress to strain, (2) correct theories for the analysis of stress and deformation resulting from either the bending or twisting of a structural member, and (3) the extension of column-buckling theory to the buckling of plates and the lateral-torsional buckling of beams.

The foregoing advances made possible the development of engineering specifications built around the *allowable-stress method* of selecting structural members. The first general specification for steel railway bridges was developed in 1905, and the first highway bridge specification in 1931. In 1923 the AISC brought out its first general specification for building construction. Under each of these specifications, the criterion for acceptable design is as follows: the calculated maximum stress, assuming elastic behavior up to anticipated maximum loads, is kept lower than a specified allowable stress. The allowable stress is intended to be less than the stress causing failure by a *factor of safety*. Unfortunately, the calculated elastic stress at failure load varies widely. A slender column or laterally unsupported beam may fail at a fraction of the yield point stress, but a very short column will reach the yield point before it fails. A statically loaded tension member may develop the ultimate tensile strength of the material, or nearly twice the yield point; but the same member, loaded and unloaded repetitively for thousands of cycles, may fail due to fatigue at a fraction of the yield point. A connection, because it yields locally, may not fail until the calculated *elastic* stress is several times the yield point; but it, too, is susceptible to fatigue failure at much lower stresses. It is evident that the true criterion of acceptability is strength—not stress—and thus, on the basis of experience and strength analyses, specified allowable stresses have had to be adjusted upward and downward over a wide range to provide a reasonably uniform index of structural strength.

During the past 60 years, and especially during the past 30, increasing attention has been given to the evaluation of the inelastic properties of materials and to the direct calculation of the ultimate strength of a member. This information is useful in improving the allowable-stress procedure, but it also permits bypassing stress calculation by using the calculated member strength as a direct basis for design. *Load-factor* design has resulted. The maximum anticipated service loads are multiplied by a load factor to yield a required strength, which must be less than the directly calculated strength. Philosophically, this is a more realistic, direct, and natural procedure. The load-factor approach has been used for many years in aircraft design, and Part 2 of the AISCS, introduced in 1961, permits it now as an acceptable alternative to allowable-stress procedures for the design of continuous frames in building structures. Although the current trend in design is to deemphasize the calculation of stress, such calculations are still essential in the design of machine parts and structural elements that must endure many load repetitions. Total resultant stresses must also be calculated in truss analysis and design.

1.5   STRUCTURAL DESIGN ECONOMY

In a competitive world, with increasing costs for materials and labor, the search for greatest design economy consistent with safety and the desired life of the structure is of major importance. Members must be shaped, arranged, and connected in ways that will provide an efficient and economical solution to the design problem, having in mind not only the cost per pound of the material itself, but also the labor costs of shop fabrication and field erection. Minimum weight is often a design goal. However, if simplicity of fabrication is sacrificed to achieve minimum weight, the overall cost may be increased. In Fig. 1.3(a) a steel beam under uniform load will be adequate in strength if it is fabricated as shown out of three segments—two end pieces, labeled (1), which weigh less per foot of length than the center section, labeled (2). But the cost of welding the three segments together may (or may not) exceed the cost of the added weight if the beam is made of a single nonwelded member having the same size as the center segment, as shown in Fig. 1.3(b).

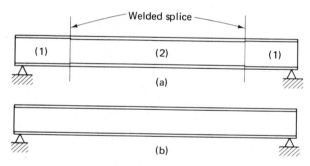

Fig. 1.3   Reducing weight may increase cost.

A similar situation may occur in plate girder design.† The use of very thin webs is made possible by vertical and (in some cases) horizontal stiffeners welded to the web. In borderline instances, the use of a thicker web, which eliminates the need for stiffeners, can result in a saving in fabrication cost even though the overall girder weight is increased.

Figure 1.4 illustrates how member arrangement can affect economy. In each trussed rectangular frame, load H is applied horizontally at the top, acting as shown, and only at A is there horizontal reaction support. In arrangement (a), only two of

†See Chapter 7.

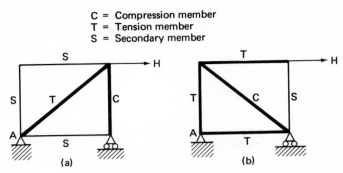

Fig. 1.4 Economy as affected by member arrangement.

the five members are directly stressed by the load. These are called *load-bearing* members and are indicated by the heavily weighted lines. But in arrangement (b), four of the five members are stressed by the load. The unstressed members would be called *secondary* members. Moreover, in arrangement (b) the compression (C) member is longer than in arrangement (a), thus using more material because of the lesser efficiency of compression members in comparison with tension (T) members. The foregoing illustrates the general principle that greatest economy results by providing the most direct path possible for the transmission of force from load point to footing.

Method of shipment can have an important bearing on economy. Connections can be made in the fabricating shop at a fraction of the cost for the same connections made during field erection. A fabricating plant built on a navigable waterway has a great advantage in building a bridge over a river accessible to the same waterway. Girders several hundred feet in length can be shop fabricated with no field splices and shipped direct to the site on barges. The same girders, if shipped by rail or truck, would need several field-splice connections, and if their overall height exceeded rail or highway clearance limitations, horizontal field splices would also be necessary.

In short-span structures, dead weight contributes but little to the stress. But as the span increases, so does the proportion of the dead-load stress to the total combined stress. Finally, when the span is so great that all the stress is due to dead load, the upper limit of span has been reached for that material and that type of structure. Thus, in long-span bridges or in tall buildings, careful attention to weight reduction and accuracy in dead-weight computations take on increasing importance. In such structures the use of high-strength steels for load-carrying members and lightweight metals for non-load-carrying elements may be advantageous.

1.6 STRUCTURAL SAFETY

Structural safety can be assured by a combination of good design, good workmanship in fabrication, and good construction methods. Unfortunately, some structural failures occur from time to time.

In design, the choice of a proper load factor for plastic design or proper unit stresses and procedures of analysis for allowable-stress design requires experience and sound engineering judgment. Questions of deterioration due to corrosion during the planned life of the structure, variation in material properties, and many other factors also need to be considered. The most rational approach to the problem of structural safety requires a statistical evaluation of the random nature of all the variables that determine the strength of the structure, on the one hand, and those that may cause it to fail (primarily, the loads) on the other hand. Then, by elementary probability theory, the risk of failure may be evaluated and the probability of its occurrence kept at an acceptable level, dependent on the importance of the structure, risk to human life, and other factors. Increasing attention is being given to this approach to safety evaluation, and statistical studies are being made of material properties, variation in strength of various types of members, and loads. Particular attention also has to be given to uncertain loads, such as those due to wind and earthquake.

Years of design experience, conditioned by both unsuccessful and successful behavior, have produced criteria that aid the choice of safe stress levels. These may not produce the most economical structure; nevertheless, the overall cumulative experience in engineering design has provided a background from which the engineer derives confidence in many particular design applications. Obviously, great skill, care, and more detailed stress analyses, possibly supplemented by laboratory tests of models or portions of a prototype structure, are needed when the designer attempts a new and venturesome type of structure.

1.7 PLANNING AND SITE EXPLORATION FOR THE SPECIFIC STRUCTURE

After the decision is made to build a steel structure to fill particular service functions, consideration is given to those factors that may influence the overall economy. If alternative sites are available, in the case of large and heavy structures, preliminary explorations are required of the various locations with a mapping of the terrain and partial preliminary study of subsurface foundation conditions by means of borings and/or open-pit excavations. Load-bearing tests may be required. If the terrain is uneven, certain functions of a building may make advantageous use of the changing elevation of the ground, and this will of course affect the overall structural layout. Other influencing factors are transportation facilities, availability of water, gas, and other utilities, drainage features, orientation with respect to prevalent winds, consideration for daylight lighting, and the general type of foundation to be required. After all these considerations and after selection of the exact site, more test borings should be made if there is any question at all about the foundation conditions or their uniformity. It may very well happen that the preliminary borings have straddled some subterranean stream location, hard strata lenses, or rock faults with associated local poor foundation conditions. In such a case, misleading estimates resulting in costly changes in design may be avoided by complete subsurface exploration. In recent years subsurface seismic surveys have proved remarkably accurate in locating bedrock and other hard layers. Such surveys are much cheaper than borings and may be used as a preliminary step covering wide areas, followed up by dry sample borings covering smaller areas selected by the results of the seismic survey. In regions subject to settlement of foundation or questionable soil capacity, undisturbed soil samples should be taken for laboratory tests of unconfined and confined compressive strength, shear strength, degree of consolidation, permeability, and so forth. If pile foundations are used, test piles may be required.

1.8 STRUCTURAL LAYOUT, DETAILS, AND DRAWINGS

After preliminary drawings have been made of the space and area requirements, in plan and in elevation, and general decisions made as to materials, type of structure, and so forth, the designer may proceed with a preliminary trial location of columns and footings.

With respect to both design and fabrication, *economy results from simplicity and duplication.* Duplication also leads to the use of standardized mass-produced elements, such as windows and doors, and has led to what is termed *modular construction.* The module is a basic space dimension that repeats itself throughout the structure and may apply to the column spacing or to smaller details. The module in building construction is frequently about 5 ft, usually a multiple of 4 in. This module is used throughout the building and applies to partitions, ceilings, lighting, windows, and so on. Columns may be four or five modules on center, as part of the modular scheme. Thus standarized spacing results in an increase in duplication and standardization of these details. Duplication in floor and roof construction will also result from the constancy of column spacing chosen in the basic modular design concept. Such duplication leads in the fabrication shop to fewer different sizes and lengths of beams in the overall steel order, and the duplication of beam and column details reduces the number of design detail drawings that are required. Repetition speeds up the work in the shop with corresponding reduction in cost.

The choice of roof and exterior wall construction involves the possibility of selecting some commercially developed standardized roof or wall product that will determine within reasonable limits the spacing of purlins and girts. However, within the range of feasible variations in purlin and girt spacing consistent with the modular layout, preliminary designs and cost estimates of various spacings should be made to determine the least weight of steel, and this will usually be of least cost as well.

Procedures for the design of main members, such as beams, columns, and tension members, are fairly simple and precisely laid down by specifications. It is in the design of the connecting details between members and their supports that the structural engineer is called upon for the greatest judgment and design skill. Poorly designed connections may lead indirectly to the failure of main members or even the entire structure. In any structure the load must be transmitted through successive connections from points of application down to the footings. The designer must follow the same sequence—for each succeeding component part of a structure must carry the accumulated dead weight of tributary components, and in preliminary design studies these weights can only be roughly estimated.

Important in the design of details is the elimination of bending or eccentricity in local elements. As one specific example, if a column is carried on top of a beam (or acts as a local beam support), the webs of the column and beam should be in alignment; but, since the major load in the column is carried in the flanges, the flanges should in turn be supported by bearing stiffeners that are directly beneath. Thus the load is transmitted from point to point throughout the structure in the most efficient manner without possibility of local failure.

Careful attention must be given in the design of structural details to the method of fabrication and erection, with proper clearances for bolting or riveting, welding electrodes and holders, or whatever else may be required for the fabrication process in question. The designer of details should visualize the complete construction operation. Attention must also be given in the details to drainage holes in pockets of exposed steel construction, because moisture and dirt should not be permitted to collect at these points of greatest incipient corrosion.

Design drawing should be complete and easy to follow. The AISC textbook *Structural Steel Detailing* provides an authoritative guide. As in many other separate aspects making up the whole of an engineering design, the *saving* of money in preparation of design drawings may result in greatly *increasing* the overall costs of the structure. If the structure is to be bid upon by private organizations, lack of sufficient detail in the design drawings may cause the bidder to add an appreciable amount for the contingencies that he must be prepared to face later when the details of connections and other framing problems are fully brought to light. If these are completely shown in the initial design drawings, the bidder will be able to give the best possible price.

1.9 FABRICATION METHODS

For greatest economy the design must be made in the light of a preliminary decision as to material and mode of fabrication. For example, the economical use of welding largely results from the introduction of continuity (in either allowable stress or plastic design) and from the elimination of the connecting pieces that would be needed in riveted or bolted construction. Although it is usually uneconomical to use both bolting and welding in the shop fabrication of a particular member, because of the double handling, the designer should consider the use of shop welding with bolted field connections. This is especially appropriate in the case of truss bridges, with high-strength bolts used for the field connections. Such connections have an excellent record for resistance to repeated load.

The use of welding requires careful and competent inspection both with regard to procedure and finished product. Both shop and field inspection of welding are important, as the quality of welds depends to a large extent on the skill, character, and endurance of the welder.

When bolting or riveting is to be used, the question arises as to whether punching of holes, subpunching with reaming, or drilling should be employed. Punching with automatic spacing equipment and repetition of members having the same punch pattern is a very economical means of preparation for bolts and rivets. However, punching damages the material locally at the edges of the holes, and such members are not as good under repeated load as are members with drilled holes. Of course, only such members as will receive large fluctuations in applied load require consideration of fatigue strength. There would be no point in subpunching and reaming—or drilling—holes for the connection of roof purlins to their truss supports, because the maximum loads are repeated a relatively few times and the stresses are minimal. In the case of shop assemblies joining several different plates or members, economy may be achieved by clamping the pieces into a single "pack" for single or multiple drilling through all pieces in one operation. Drilling provides smooth edges of holes and the best possible resistance to repeated load.

1.10 CONSTRUCTION METHODS

Structural designs should be prepared with ample consideration of the manner and facility with which field erection can proceed. The arrangement, number, type, and location of field splices and connections should be planned so as to avoid unnecessary duplication of construction equipment and provide the simplest possible erection plan with a minimum of field work. Connections should be arranged to facilitate field assembly. Careful design planning in relation to construction will minimize the total cost of the project. In important large projects a definite erection plan should be presented, but the contractor should have freedom to exercise his own ingenuity through alternative schemes that meet the approval of the owner.

In one particular sense, proper construction methods have a special relation to overall economy, since it is during construction that failures in engineering structures most often occur. One cause of failure occurs during lifting operations of trusses and girders. Members of trusses that normally are in tension, or the lower flange of a plate girder, which is normally in tension, may be placed in compression with consequent possible buckling failures.

Even after the main frames and members are successfully placed in the structure, failures have occasionally occurred because of the haste with which construction of main framing has proceeded without attention to the cross bracing that may be planned for the final structure in the planes of the walls and roof. After permanent bracing, roof, and walls are in place, the wind load resistance of the structure will be greatly increased. In the case of very long plate girders used in bridge construction, experienced contractors provide special horizontal temporary truss systems fixed to the plate girders for use only during erection. Although erection is normally the responsibility of the steel contractor, the design engineer can help in complex cases by scheduling the bracing that must be supplied as the construction is in progress. Alternatively, the contractor may be required to submit erection procedure plans to the engineer for approval.

In summary, it may be said that construction failures are usually caused by lack of three-dimensional or "space frame" stability and that many more failures occur during erection than during service of the finished structures.

1.11 SERVICE AND MAINTENANCE REQUIREMENTS

The engineer, together with the architect and special consultants on such matters as heating, lighting, and ventilation, should give careful attention to the way in which the utility of the structure may be affected by the engineering design. Especially in an industrial building, expediency of structural design must sometimes take second place to the service functions of the structure.

Inadequate initial planning regarding the service which the structure is to perform will inevitably result in revisions in layout and corresponding costly design and material order changes before the structure is complete. It is obvious also that the locations of the electric wiring, heating ducts, and other service ducts for water, gas, chemicals, and so forth, as well as the locations of all special items of equipment, must be carefully predetermined, as all affect and are affected by the structural design.

Another service requirement of concern to the engineer is the desired life of the structure, together with consideration as to any special problems of corrosion that may exist due to atmospheric conditions, humidity, and so on. By proper design the engineer should avoid pockets where dirt and water may collect to cause corrosion, and should provide access to all parts of the structure that will require repeated painting and inspection during its life. Under adverse conditions when maintenance cannot be assured, an extra thickness of metal may be furnished to allow for corrosion. Special corrosion-resistant steels are available, and another alternative is the use of *weathering steels*, which require no paint and develop a surface oxide that resists corrosion and presents a pleasing burnt-brown color.

During the useful life of an industrial plant, changes in processing procedures or even a complete change in use may come about. Thus, along with all necessary attention to special service requirements, an effort should be made to incorporate flexibility with respect to possible future alterations. The use of temporary interior partition walls is an example of such flexibility with respect to future change.

A structure should be designed to provide a life consistent with the buyer's wishes. A structure that is to last 100 years will be of quite different construction than one designed to survive as a temporary structure for only a few years. The choice of materials used in construction may be affected; but even if both structures were of steel, there would need to be different consideration given to the problems of permissible stress, load evaluation, corrosion, painting, and other matters of upkeep for the two different life expectancies. The use of closed tube or box sections may materially reduce painting maintenance costs and be more justified in a long-lived structure than in a temporary one. Consideration under this category should also be given to fireproofing and fire protection. The difference in cost of fire insurance over the life of the structure must be weighed against the difference in initial cost between various degrees of fireproofing, assuming, however, that safety to human life is not an overriding consideration.

REFERENCES

1.1. *Specifications for the Design, Fabrication and Erection of Structural Steel for Buildings* with Supplements No. 1 and No. 2 of 1970 and 1971. American Institute of Steel Construction (1969).

1.2. *Manual of Steel Construction,*† 7th ed. American Institute of Steel Construction (1970) (includes Ref. 1.1).

1.3. *Structural Steel Detailing.* American Institute of Steel Construction (1971).

1.4. *Specification for the Design of Cold-Formed Steel Structural Members.* American Iron and Steel Institute (1968).

1.5. WINTER, GEORGE. *Commentary on the 1968 Edition of the Specification for the Design of Cold-Formed Steel Structural Members.* American Iron and Steel Institute (1970).

1.6. JOHNSTON, B. G., ed., *The Column Research Council Guide to Design Criteria for Metal Compression Members,* 2nd ed. John Wiley & Sons, Inc., New York (1966).

1.7. BROCKENBROUGH, R. L., and B. G. JOHNSTON. *The USS Steel Design Manual.* The United States Steel Corporation (1968).

---

†A mandatory supplement for the complete use of this book. Reference 1.1 will be referred to herein simply as AISCS; Reference 1.2 will be referred to as AISCM.

2

TENSION MEMBERS

2.1 INTRODUCTION

The most efficient way to use structural steel is in a tension member, that is, one that transmits "pull" between two points in a structure. Of course, if under certain load conditions the stress reverses in the member and becomes compression, the member must be designed both as a tension member and a column, and the efficiency is lost.

To make all the material in the tension member fully effective, the end connections must be designed to be stronger than the body of the member. If overloaded to failure, such a tension member will not only reach the yield stress but go above this level up to the ultimate strength of the material. In so doing it can absorb a great deal more energy per pound of material than any other type member. This is an important consideration if impact or dynamic loads are a possibility. Beams and columns do not utilize material at full efficiency for two reasons: (1) Metal failure is localized at highly stressed locations and (2) some type of buckling failure always occurs at or below the yield stress, and the ultimate tensile strength of the material can never be reached.

Four types of tension member that can achieve high efficiency as previously described are illustrated in Fig. 2.1, showing: (a) The wire rope or cable with socketed ends in which the use of cold-drawn steel wires having tensile strengths up to 150 ksi (or more, in special applications) provide the greatest strength–weight ratio available in the use of steel, (b) the simple round rod with threaded upset ends, (c) the eyebar, with forged ends for pin connections that are stronger than the body of the bar, and (d) the pin-connected plate with welded reinforcing plates at the ends.

In contrast to the foregoing, a tension member that can fail within its end connection before yielding in the body of the member will absorb little energy before

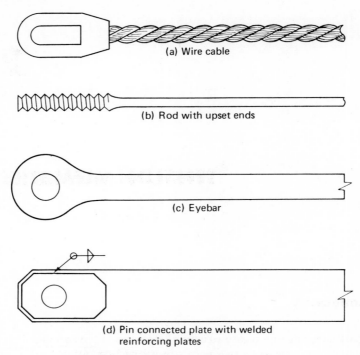

(a) Wire cable

(b) Rod with upset ends

(c) Eyebar

(d) Pin connected plate with welded
reinforcing plates

Fig. 2.1 High efficiency tension members.

failure—possibly less than 1 per cent of the capacity it would have with uniform yielding throughout its length. Regardless of where failure under static load might occur, the tension member and its end connections should be designed to guard against fatigue failure if alternate loading and unloading is to be expected for a large number of repetitions.

Because of their efficiency, and because buckling is not a problem, tension members make more advantageous use of the higher-strength steels than any other type member.

No structural member is perfectly straight, and an intended axial force will never act precisely along the longitudinal axis. As a result, there are always "accidental" bending moments in such a member. In a column, as illustrated in Fig. 2.2(a), these bending moments cause added deflection which further increases or "amplifies" both the deflection and the bending moment caused thereby, equal to the product of the axial load and the deflection.

In an initially curved and eccentrically loaded tension member [Fig. 2.2(b)] the member tends to straighten, and the bending moments are reduced everywhere except at the end. Thus, for very small accidental curvatures and end eccentricities, the additional tension stress induced by bending can usually be neglected unless design for repeated load is required.

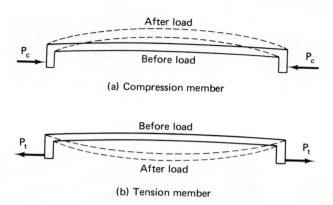

(a) Compression member

(b) Tension member

Fig. 2.2 Deflection of eccentrically loaded compression and tension members.

2.2 TYPES OF TENSION MEMBERS

Four efficient tension-member types have been illustrated in Fig. 2.1. In addition, structural shapes and builtup members may be used, especially in trusses where tension and compression members must frame into a common joint, as shown in Fig. 2.3.

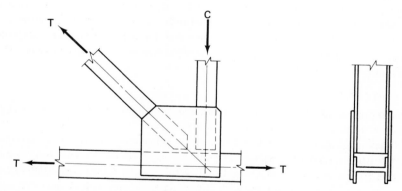

Fig. 2.3 Tension members (T) and compression member (C) entering lower chord joint of a truss.

(a) *Wire Ropes and Cables*

A cable is defined as a flexible tension member consisting of one or more groups of wires, strands, or ropes. A strand is formed by wires laid helically about a center wire so as to produce a symmetrical section; a wire rope is a plurality of strands laid

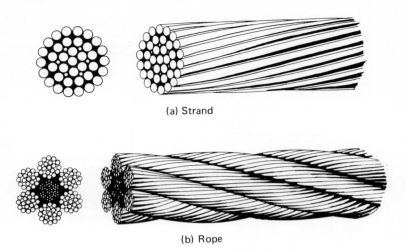

(a) Strand

(b) Rope

Fig. 2.4 Wire strand and wire rope, from Ref. 2.1. (U.S.S. Corporation.)

helically around a central core that is composed of a strand, fabric core, or another wire strand, as illustrated in Fig. 2.4, from Reference 2.1.

Wire cables are finding increasing use in structural steel design, and have been used both as primary and secondary supporting members in a wide variety of structures, including suspension bridges, prestressed concrete members, guyed towers, and wide-span roof structures. In roof construction the cables may radiate outward from a central tower, or may run radially inward from an outer compression ring, as illustrated in Fig. 2.5. The major U.S. steel producers have available manuals, such as Reference 2.1, that provide extensive design information and illustrations of the use of cables in roof structures.

(b) *Rods and Bars*

The simplest tension member is the square or round rod. Round bars with threaded ends are less costly than bars with upset ends† [Fig. 2.1.(b)], but have certain disadvantages. Failure under impact overload or repeated load is apt to occur in the threaded portion. Bars with upset ends yield throughout their entire length and are recommended for the design of cross bracing for simple tower structures in earthquake regions. Large-diameter bars with threaded ends should be used with caution, because the lateral contraction in the diameter of the bar as yielding commences in the threaded portion may result in sufficient loss of thread-bearing area to result in failure by stripping of the threads before fully developing the maximum desired strength.

---

†Upset ends were originally formed by forging. Currently, the threaded parts may be made from a larger diameter rod than the main plain portion and the three segments are then butt welded together.

To guard against loosening after overloading, provision may be made for tightening at the ends of the member or by means of a turnbuckle between the ends of a two-piece member.

Round bars are frequently grouted into holes in rock formations to contain tunnel liners or retaining walls. They are also useful in reducing and restraining movement, such as a cracked machinery pedestal or spreading walls in old masonry structures.

In designing a rod with upset ends, the average stress on the area at the root of the thread should be less than the stress in the body of the bar. This will ensure yielding in the body of the bar if there is severe overload, as in an earthquake. If there is misalignment or bending in a tie rod, the use of upset ends offers the additional advantage that any added stress due to bending is greatest in the main body of the bar, which is most flexible and able to adjust to such a condition.

Fig. 2.5 Wire strand roof support system. (Am. Inst. of Steel Construction.)

Areas of rods and bars are tabulated in the AISCM from pages 1-110 to 1-111. Dimensions of threads, turnbuckles, clevises, and sleeve nuts are given on pages 4-125 to 4-128 and are listed for a wide variety of rod diameters.

(c) *Eyebars and Pin-Connected Plates*

Eyebars and pin-connected plates [Figs. 2.1 (c) and (d)] are used in a variety of special situations. Examples include the transfer of tensile load from a wire rope or cable to a structural steel assemblage or to an anchorage, as in the case of a suspension bridge. The use of eyebars as tension members in a modern long span bridge is illustrated in Fig. 2.6.

Fig. 2.6 Eyebar tension members in the Brent Spence Bridge over the Ohio River.

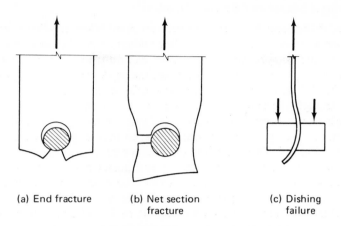

(a) End fracture   (b) Net section   (c) Dishing
                              fracture   failure

Fig. 2.7   Various failure modes of a pin-connected plate.

If failure were to occur in the head of the eyebar at the connection, tests have demonstrated that it would be one of the following types:

1. Fracture behind the pin in a direction parallel to the axis of the bar. This type of failure will occur if insufficient edge distance behind the pin is provided [Fig. 2.7(a)].

2. Failure in the net section through the pin transverse to the axis of the bar. This type of failure will occur if the gross area of the main section of the bar is equal to or greater than the net section through the pin hole [Fig. 2.7(b)].

3. Failure by *dishing*. This is an inelastic lateral stability type of failure, which will occur if the width–thickness ratio behind the pin is too great. Dishing failure is akin to the lateral instability of a short deep beam [Fig. 2.7(c)].

Since it is desirable to ensure that general yielding and ultimate failure occur in the main body of the bar rather than at the end, all specifications provide dimensional requirements that prevent failure of the types enumerated above. However, if there are several eyebars coming together or connecting at the same pin with packing and external nuts to prevent spreading of the package, there will be no need to restrict the width–thickness ratio of the eyebar, since dishing will be prevented by the lateral constraints provided. This is the usual design situation. The required proportion for eyebars is similar in the various specifications.

In general, connection details are treated in Chapter 6; but in the case of eyebars or pin-connected plates, the design of end-connecting details, except for the proportioning of the pin itself, is related to the design of the whole member and will be covered in this chapter. The AISCS, Sec. 1.14.6, specifies proportions that provide for a balanced design with respect to the various possible modes of failure. It should be studied in detail by reference to Ex. 2.1.

(d) *Structural Shapes and Built-up Members*

Structural shapes and built-up members are used when rigidity is required in a tension member, to resist small lateral loads, or when reversal of load may subject the member to alternate compression and tension, as in a truss diagonal near the center of a span. The most commonly used shapes are the angle, tee, and W, S, or M shapes, as shown in Fig. 2.8. For exposed use, to minimize wind load, the pipe section may be favored, although the end connections present a problem in truss construction. Built-up members are formed by connecting two or more structural shapes with separators, battens, lacing, or continuous plates, so that they will work together as a unit, as shown in Fig. 2.8. The angle and channel members, as shown, may be used in single-plane truss construction connected to end gusset plates with rivets, bolts, or welds. The built-up open-box shape, as shown (the dashed lines indicate lacing or battens), is suitable for double-plane truss construction and is conveniently bolted, riveted, or welded between two gusset plates at each end connection. W, M, or S shapes are especially suited to double-plane welded truss construction.

The tension members must, if riveted or bolted together, be designed on the basis of the net cross-sectional area, which is defined by appropriate specifications. The average or nominal stress at the net section is used as the basis for design. As a result of many years of experience, this has proved to be a safe practice.

Although end connections are of paramount importance in the design of a complete tension member, this topic, except for the eyebar, will be covered in Chapter 6.

Fig. 2.8 Structural shapes as tension members.

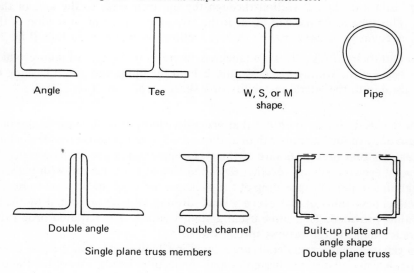

| Angle | Tee | W, S, or M shape | Pipe |

Double angle   Double channel   Built-up plate and angle shape

Single plane truss members   Double plane truss

2.3 ALLOWABLE TENSILE STRESS

Axially loaded tension members are proportioned so that the nominal, or average, stress shall not exceed the specified allowable tensile stress as defined by the specification as a proportionate part of either the yield stress or ultimate tensile strength, whichever is critical. The nominal tensile stress is simply the anticipated axial design load P divided by the net cross-sectional area A of the member:

$$f_t = \frac{P}{A}$$

Although an effort should be made to reduce local concentration of stress due to changes in section, particularly at connecting welds, by use of smoothly tapered and gradual transitions, the local stresses are not usually added to the "nominal" stress. Tests to failure have shown that regions of local yield in a well-designed and properly fabricated tension member do not prevent the entire cross section from reaching the yield point and beyond, thus developing the full strength of the member before failure.

For buildings, the allowable tensile stress is specified by the AISCS, Sec. 1.5. 1.1, as follows:

1. Allowable tensile stress F_t on the net section of the tension member, except at pin holes:

$$F_t = 0.6F_y \leq 0.5F_{ts}{}^*$$

where F_{ts} is the minimum tensile strength of the steel. F_t (in kips per square inch) is tabulated as follows for the various yield stresses:

F_y	36	42	45	50	55	60	65	90	100
F_t	22	25.2	27	30	33	36	39	52.5*	57.5*

2. Allowable tensile stress F_t on the net section at pin holes in eyebars, pin-connected plates or built-up members:

$$F_t = 0.45F_y$$

and is tabulated as follows:

F_y	36	42	45	50	55	60	65	90	100
F_t	16.2	19	20.3	22.5	24.8	27	29.3	40.5	45

A nonmandatory clause, Sec. 1.8.4. of the AISCS, page 5-25, recommends a maximum slenderness ratio (l/r) of 240 for main members and 300 for bracing and other secondary members. Rods are excepted from these limitations.

2.4 DESIGN FOR REPEATED LOAD

When load is repeatedly applied and removed, with the number of repetitions running into the many thousands or up into the millions, metal may develop cracks that eventually may spread to the point where they cause *fatigue failure* of the member. Fatigue cracks are most apt to occur when the repeated load is primarily tension. Local stress concentrations increase the susceptibility to fatigue failure. Such concentrations may be due to poorly made welds, small holes, and rough or damaged edges resulting from the fabrication processes of shearing or poor-quality oxygen cutting. The fatigue strength of the higher-strength steels has not been shown to be appreciably greater than the commonly used A36 grade structural steel with a yield stress of 36 ksi.

In 1969 the AISCS introduced a new and simplified approach to design for repeated load, as presented in Appendix B to the AISCS, pages 5-107 to 5-113. This appendix should be studied in detail, and the following will simply outline its coverage.

The unique feature of the AISCS approach to repeated load design is its use of the expected *stress range* as the controlling design criterion. Stress range is the algebraic difference between maximum and minimum stress to be expected in any one cycle of loading. Thus the following two cases each have the same stress range of 16 ksi:

Maximum stress	Minimum stress
20 ksi tension	4 ksi tension
4 ksi tension	−12 ksi compression

The permissible stress range is a function of (1) loading condition and (2) stress category.

Loading conditions are defined and tabulated in Table B1, page 5-107, of the AISCS, and are determined according to the anticipated number of loading cycles to be used as a basis for design. If less than 20,000 cycles, no consideration need be given to repeated load, but at successive lower limits of 20,000, 100,000, 500,000, and 2,000,000 cycles, respectively, load conditions 1, 2, 3, and 4 are established.

The stress categories, ranging from A to G with increasing severity of the local stress raiser, are tabulated and defined in Table B2 of the AISCS, pages 5-108 to 5-111.

After establishing the loading condition and the stress category, the allowable stress range is read from AISCS, Table B3. No differentiation is made for variation of steel yield stress except for A514 steel, in category A, alone, where there are slightly greater stress ranges. It should also be noted that Eq. (B1), AISCS, page 5-107, provides for an increased allowable stress range when there is stress reversal. The amount of increase is based on the increase in ratio of maximum compressive stress to maximum tension stress.

Members in conventional buildings usually do not need to be designed for repeated load because the number of repetitions of maximum load is usually less than 20,000. However, crane runway girders and supporting structures for machinery and equipment do require the consideration of fatigue.

2.5 FLOW CHART

Flow Chart 2.1. *Tension member selection.*

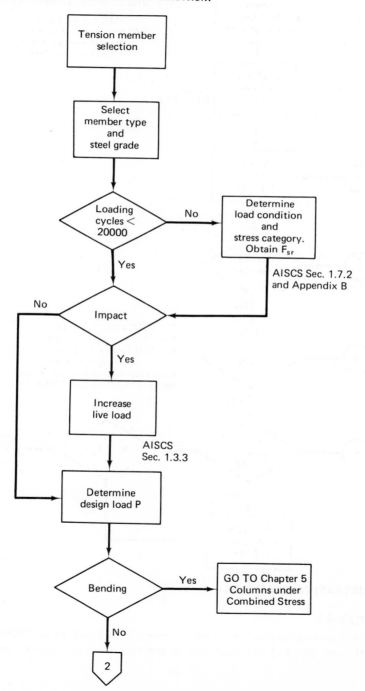

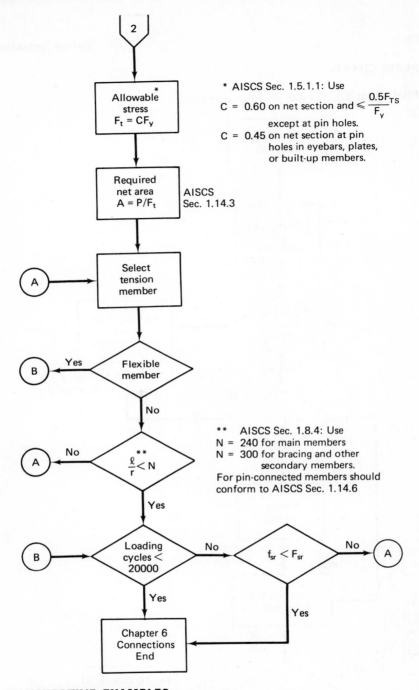

* AISCS Sec. 1.5.1.1: Use

$C = 0.60$ on net section and $\leq \dfrac{0.5F_{TS}}{F_y}$

except at pin holes.

$C = 0.45$ on net section at pin holes in eyebars, plates, or built-up members.

** AISCS Sec. 1.8.4: Use
N = 240 for main members
N = 300 for bracing and other secondary members.
For pin-connected members should conform to AISCS Sec. 1.14.6

2.6 ILLUSTRATIVE EXAMPLES

Example 2.1

A floorbeam suspender is stressed in tension by a dead load of 30 kips and a live load of 40 kips. The full live load will be repeated less than 20,000 times. Select a round steel rod, using upset ends, to satisfy the AISCS. Use quenched and tempered alloy steel with $F_y = 100$ ksi.

Solution

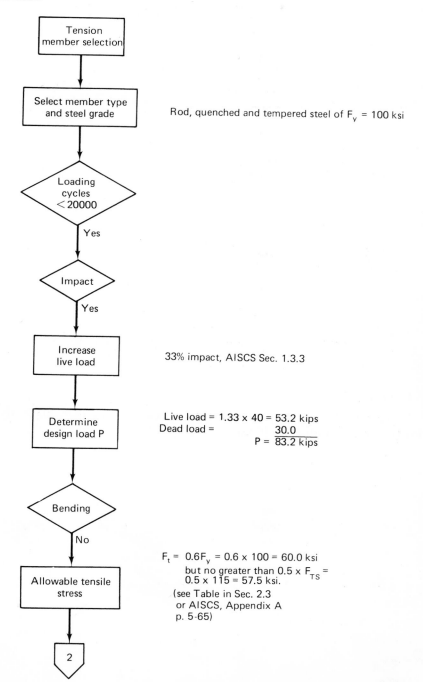

Rod, quenched and tempered steel of $F_y = 100$ ksi

33% impact, AISCS Sec. 1.3.3

Live load = 1.33 x 40 = 53.2 kips
Dead load =   30.0
                      P = 83.2 kips

$F_t = 0.6F_y = 0.6 \times 100 = 60.0$ ksi
but no greater than $0.5 \times F_{TS} = 0.5 \times 115 = 57.5$ ksi.
(see Table in Sec. 2.3
or AISCS, Appendix A
p. 5-65)

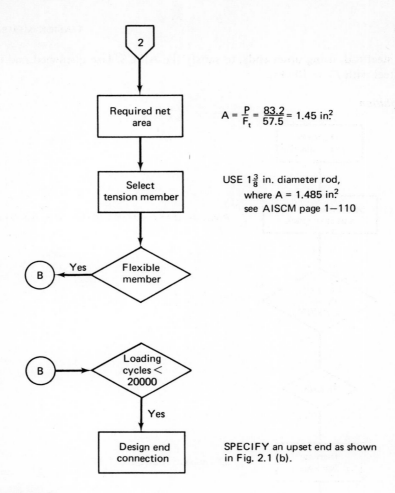

$$A = \frac{P}{F_t} = \frac{83.2}{57.5} = 1.45 \text{ in.}^2$$

USE $1\frac{3}{8}$ in. diameter rod,
where A = 1.485 in.^{2}
see AISCM page 1–110

SPECIFY an upset end as shown
in Fig. 2.1 (b).

Example 2.2

Design an eyebar to carry a tensile load of 600 kips (less than 20,000 repetition of load). Use steel with $F_y = 50$ ksi.

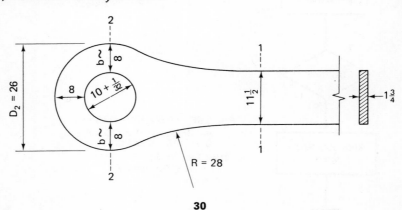

Given: P (ten.) $= 600\,k$ and $F_y = 50$ ksi.

Solution

(1.5.1.1)† $F_{t_1} = 0.6(50) = 30$ ksi $F_{t_2} = 0.45(50) = 22.5$ ksi

$A_1 = \dfrac{600}{30} = 20$ sq.in.   Use PL $1\frac{3}{4} \times 11\frac{1}{2}$ ($A_1 = 20.12$ sq.in.)

(1.14.6)† $\left(\dfrac{b}{t}\right)_1 = \dfrac{11.5}{1.75} = 6.57 < 8$ OK

A_2 (min) $= \dfrac{600}{22.5} = 26.6$ sq.in.

or $1.33 \times A_1 = 26.8$ sq.in. ← Governs

b at 2–2 $= \dfrac{26.8}{2 \times 1.75} = 7.67$ in.   Use 8 in.(approx.)

A_2(act.) $= 2 \times 8 \times 1.75 = 28.0$ sq.in. $< 1.5\,A_1 = 30.2$ sq.in.   OK

diam. pin $\geq \frac{7}{8}$ (11.5) $= 10.08$ sq.in.   Use 10 in. pin

diam. hole $= 10 + \frac{1}{32} = 10\frac{1}{32}$

$D_2 = 10\frac{1}{32} + 2 \times 8 = 26\frac{1}{32}$ Use 26 in. ($b \approx 8$ in.)

$R > D_2 = 26\frac{1}{32}$ Use 28 in.

(1.5.1.5.1)†   Check bearing

$F_p = 0.9F_y = 0.9 \times 50 = 45$ ksi

$f_p = \dfrac{600}{10 \times 1.75} = 34.3$ ksi < 45 OK

Example 2.3

A tension member of a roof truss has a length of 25 ft and is stressed in tension by a dead load of 40 kips and a live load of 60 kips. The tension member is a main member and needs some amount of rigidity. Select a single structural tee to satisfy the AISCS. Use A36 steel.

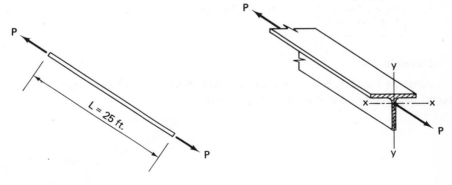

† AISCS sections.

Solution

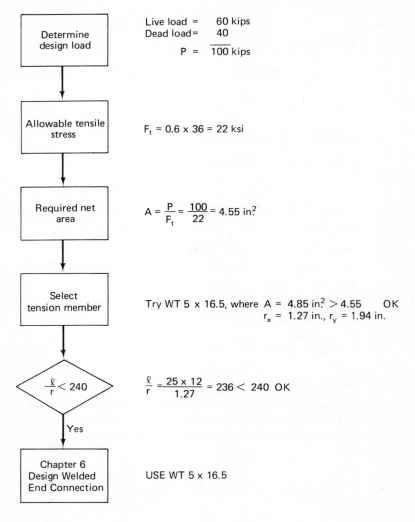

Determine design load	Live load = 60 kips Dead load = 40 P = $\overline{100}$ kips
Allowable tensile stress	$F_t = 0.6 \times 36 = 22$ ksi
Required net area	$A = \dfrac{P}{F_t} = \dfrac{100}{22} = 4.55$ in.^{2}
Select tension member	Try WT 5 × 16.5, where A = 4.85 in.^{2} > 4.55 OK r_x = 1.27 in., r_y = 1.94 in.
$\dfrac{\ell}{r} < 240$	$\dfrac{\ell}{r} = \dfrac{25 \times 12}{1.27} = 236 < 240$ OK
Yes	
Chapter 6 Design Welded End Connection	USE WT 5 × 16.5

Example 2.4

Same as Ex. 2.3, but an additional axial tension of 45 kips produced by wind should be considered if it governs the design.

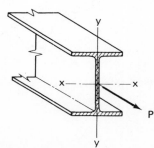

Solution

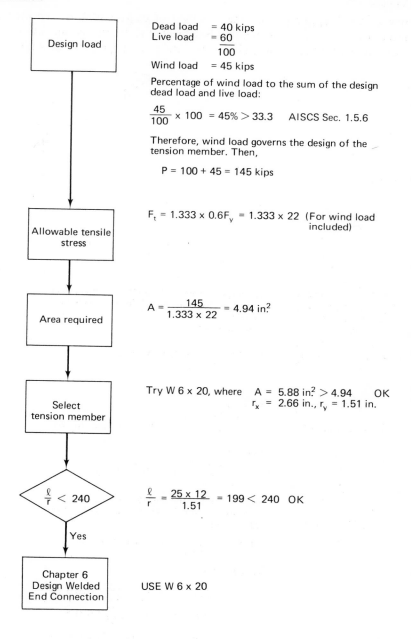

Design load

Dead load   = 40 kips
Live load   = 60
                      ——
                      100
Wind load   = 45 kips

Percentage of wind load to the sum of the design
dead load and live load:

$\frac{45}{100} \times 100 = 45\% > 33.3$ AISCS Sec. 1.5.6

Therefore, wind load governs the design of the
tension member. Then,

$P = 100 + 45 = 145$ kips

Allowable tensile
stress

$F_t = 1.333 \times 0.6F_y = 1.333 \times 22$ (For wind load
                                                                 included)

Area required

$A = \frac{145}{1.333 \times 22} = 4.94$ in.^{2}

Select
tension member

Try W 6 x 20, where $A = 5.88$ in.$^2 > 4.94$ OK
                                     $r_x = 2.66$ in., $r_y = 1.51$ in.

$\frac{\ell}{r} < 240$

$\frac{\ell}{r} = \frac{25 \times 12}{1.51} = 199 < 240$ OK

Yes

Chapter 6
Design Welded
End Connection

USE W 6 x 20

Example 2.5

Same as Ex. 2.3, except that the live load of 60 kips may be repeated 30,000 times and during each cycle the member is subjected to 10-kips compression. Welded end connection with fillet welds is similar to case 17 on page 5-113 of AISCS.

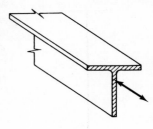

Solution

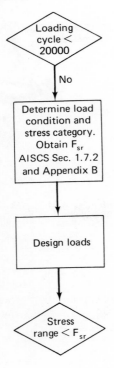

Loading condition 1 for 30,000 loading cycles.
Illustrative Example no. 17 for fillet welded connection.
Stress category E

Therefore, allowable stress range
$$F_{sr} = 17 \text{ ksi}$$

Tension:
$$\begin{array}{ll} \text{Dead load} & +40 \\ \text{Live load} & +60 \end{array} \quad P_t = +100 \text{ kips (max.)}$$

Compression:
$$\begin{array}{ll} \text{Dead load} & +40 \\ \text{Live load} & -10 \end{array} \quad P_t = +30 \text{ kips (min.)}$$

W T 5 x 16.5 is the selected tension member as Example 2.3 for the design dead and live loads of 100 kips.

Then check stress range for this fatigue loading condition:

Actual stresses:

$$\text{Max.} \quad f_t = \frac{P_t}{A} = \frac{100}{4.85} = 20.6 \text{ ksi}$$

$$\text{Min.} \quad f_t = \frac{P_t}{A} = \frac{30}{4.85} = 6.18 \text{ ksi}$$

$$\text{Actual stress range} = 20.6 - 6.18$$
$$= 14.42 \text{ ksi} < F_{sr} \quad \text{OK}$$

USE W T 5 x 16.5 (Same as Example 2.3)

PROBLEMS†

2.1. In place of the upset round rod chosen in Example 2.1, select a rectangular bar for the same design requirements. Assume welded end connections so as to make the full cross section of the bar available as net section in tension.

2.2. Same as Problem 2.1, but with high-strength bolted end connections, using a single line of 3/4 in.-diameter bolts. Refer to AISCS, Secs. 1.14.3 and 1.14.5, for guidance in determining the net section of the bar.

2.3. Redesign an eyebar to meet the requirements of Example 2.2 but with the steel yield point changed from 50 to 42 ksi.

2.4. Design a tension member to meet the requirements of Example 2.3, but with the structural tee section replaced by two angles, with long legs back to back. Long legs are to be separated 3/8 in. for end connections to gusset plates. Make use of information on pages 1-74 to 1-79 of AISCM and try to balance l/r ratios about xx and yy axes. In determining net section, assume a single line of holes for 7/8 in.-diameter bolts through long sides of angles.

2.5. Same as Problem 2.4, except that the live load of 60 kips may be repeated 600,000 times, and during each cycle the member may be subjected to a compression of 35 kips. The high-strength bolted end connection may be assumed to be equivalent, in fatigue resistance, to Case 8, page 5-112 of AISCS. Example 2.5 should be studied as a guide.

2.6. Same as Problem 2.4, but with the addition of a 40 kip tension force due to wind. Refer to Example 2.4.

2.7. The 1-3/8 Φ bar was chosen in Example 2.1 without consideration of repeated load; that is, the repetitions of load were assumed to be less than 20,000. Ruling out the possibility of failure in the end threads, show that the bar is capable of withstanding 60,000 repetitions of live load (without impact). For repeated load, the round bar may be assumed as equivalent to Case 1, page 5-112 of AISCS.

2.8. Referring again to Example 2.1, assume that the thread root area in the upset end of the rod is 1.74 in.². Assume that this puts the end in repeated load Category G of Table B3, page 5-111 of AISCS. Is the threaded end adequate for 60,000 repetitions of live load, alone, excluding impact? If not, what change would you recommend? Such a change could involve a different material, different type of member, or a different end connection.

REFERENCES

2.1. SCALZI, J. V., W. PODOLNY, Jr., and W. C. TENG. *Design Fundamentals of Cable Roof Structures.* U.S. Steel Corporation (1969).

†All problems in this and succeeding chapters are to be done in accord with AISCS. Note that the term "example" refers always to the illustrative examples in the body of the chapter and the term "problem" refers to problems at the end of the chapters. Shapes and bars chosen in problem solutions should be "available" as designated by Tables 1 and 2 of AISCM, pages 1-6 and 1-7.

3

BEAMS

3.1 INTRODUCTION

Beams support loads that are applied at right angles (transverse) to the longitudinal axis of the beam. Such loads are usually caused by the downward pull of gravity, as illustrated by the loads labeled P in Fig. 3.1(a). The beam carries the loads to its supports, which may consist of the bearing walls, columns, or other beams into which it frames. At the supports the upward "reactions" have a total magnitude equal to the weight of the beam plus the applied loads P. Since the weight of the beam is not known until after it is designed, the structural steel design of a building frame starts at the top—at the roof—and the dead weight of each structural member is added in after it is determined as the designer proceeds downward.

Imagine a free-body diagram of the left portion of the beam [Fig. 3.1(b)] with bending moment (M_R) and the shear (V) necessary at the cut section to provide static equilibrium. The problem of beam design consists mainly in providing enough bending strength and enough shear strength at every location in the span. For short spans, it is most economical to use a single-beam cross section throughout the span, and in such case only the maximum values of bending moment and shear need to be determined.

A *simple beam* [Fig. 3.1(a)] is supported vertically at each end with little or no rotational restraint, and downward loads cause positive bending moment throughout the span. The top part of the beam shortens, due to compression, and the bottom part of the beam lengthens, due to tension [Fig. 3.1(d)]. The most common rolled steel beam cross section, shown in Fig. 3.1(c), is called the W section, with much of the material in the top and bottom flange, where it is most effective in resisting bending moment. The web of the beam supplies most of the shear resistance and in so doing is slightly distorted, as shown in Fig. 3.1(e). The contribution of this distortion to beam deflection is usually neglected. The bending moment causes curvature of the beam axis, concave upward, as shown in Fig. 3.1(d) for positive moment, concave downward for negative moment. The deflection of beams is calculated on the assumption that it is entirely caused by the curvature due to bending moment.

Standard AISC nomenclature pertaining to the W (wide-flange) hot-rolled steel beams is illustrated in Fig. 3.2.

The reader should gain a familiarity with information in the AISC relative to rolled sections, reading the descriptive material and scanning the tabular material in the following AISCM locations:

Pages 1-4 to 1-10:	Discussion and tabular material relative to shape selection, designation, dimensions, availability, size groupings, principal producers, and proper manner of shape designation.
Pages 1-11 to 1-25:	Dimensions of shape cross sections for detailing.
Pages 1-27 to 1-61:	Properties of shape cross sections for use in design calculations.

Fig. 3.1 Simple beam behavior.

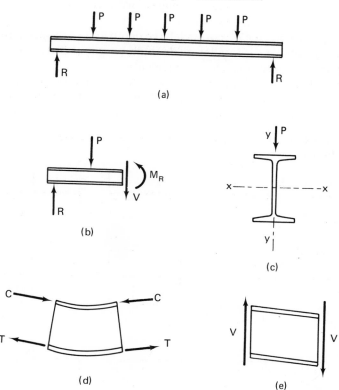

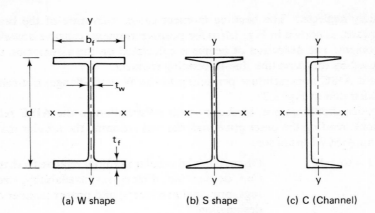

(a) W shape   (b) S shape   (c) C (Channel)

Fig. 3.2 Nomenclature pertaining to a beam cross section where: $x - x$ = principal major axis, strong axis of bending, weak plane; $y - y$ = principal minor axis, weak axis of bending, strong plane.

Pages 1-123 to 1-134:   Standard mill practice in the rolling, cambering, and cutting of shapes, with corresponding dimensional tolerances.

A *plate girder* (see Chapter 7) is of such large depth and span that a rolled beam is not economically suitable—it is tailor made (built up out of plate material) to suit the particular span, clearance, and load requirements.

It is assumed that the reader is familiar with the calculation of shears and moments, with the drawing of shear and moment diagrams, and with the usual designation of support conditions. Various cases are illustrated in Fig. 3.3. At the top the

Fig. 3.3 Load, shear, and moment diagrams for various beam situations.

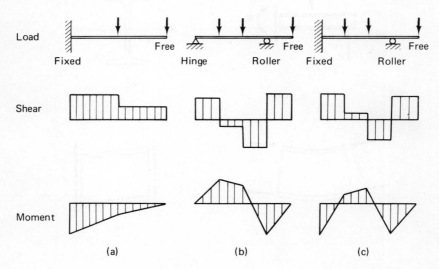

loads and supports are shown for (a) a cantilever beam, (b) a simple beam with a cantilever overhang at the right end, and (c) a beam fixed at the left end and the same as (b) at the right end. In (c) the shears and moments between the left (fixed) end and the simple support are *statically indeterminate;* that is, they cannot be determined by simple statics. Shear diagrams are shown on the second line, moment diagrams on the third. Although the calculation of shears and moments will be included in many of the illustrative examples, reference should be made to a text on strength of materials or elementary structural theory for additional information on these topics. Shear and moment diagrams for a variety of loading conditions will be found in AISCM, pages 2-198 to 2-211. For uniform, or distributed, loads the shear and moment diagrams are similar to those shown in Fig. 3.3, but the shear, since it changes with load, is a sloping line instead of a horizontal one, and the moment diagram is a continuous curve between reactions. The reader should review the mathematical relationships between load, shear, and bending moment as found in his reference text on strength of materials.

Beams usually are framed with other beams, or with a floor slab, as shown in Fig. 3.4, so that the beam cannot move sideways and the beam is forced to deflect vertically in the strong $(y - y)$ plane (see Fig. 3.2).

Fig. 3.4 Beams framed to column and supporting permanent metal forms for floor slab. (Bethlehem Steel Corp.)

Whenever a beam deflects in the plane in which it is loaded, the simple theory of bending† may be used. The condition may be forced, as previously mentioned, or it can occur naturally if the plane of the loads contains a principal axis of the cross section. However, if the load is in the strong $(y - y)$ plane (see Fig. 3.2), the beam may need lateral support to prevent it from buckling sideways; alternatively, the specifications provide for reduced allowable loads if lateral support does not meet certain minimal requirements. If loaded in the weak $(x - x)$ plane (see Fig. 3.2), lateral buckling is no problem. Sections that lack two axes of symmetry may require more positive lateral supports than does the W section when they are loaded in the usual manner, as illustrated in Fig. 3.5. For example, the laterally unsupported channel member will twist if loaded through the centroidal axis, as shown in Fig. 3.5(b), and requires restraint against both twist and lateral buckling. The zee section does not twist but deflects at an angle to the plane of the loads unless supported as shown in Fig. 3.5(c). An angle loaded as shown in Fig. 3.5(d) must be supported against both twist and lateral deflection. It is also important to recognize that if the zee or angle section is used without lateral support, the stress due to bending cannot be calculated by the simple beam formula. Where lateral support is needed to prevent only lateral buckling [Fig. 3.5(a)] there is no calculable stress in the lateral supports. In cases (b), (c), and (d), however, there is a calculable stress in the lateral supporting members, and thus a more clearly defined design problem exists.

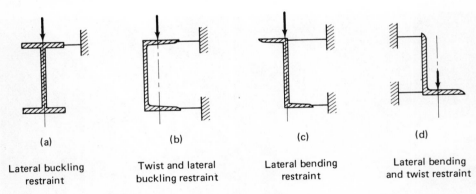

(a)	(b)	(c)	(d)
Lateral buckling	Twist and lateral	Lateral bending	Lateral bending
restraint	buckling restraint	restraint	and twist restraint

Fig. 3.5 Type of lateral restraint required to permit beam selection by simple bending theory.

Most beams are designed by simple bending theory, and the design process involves the calculation of the maximum bending moment and the selection of a beam having an equal or greater bending moment resistance. The selection is then checked for maximum shear capacity, and the end connections or bearing support details are designed. A deflection check may also be required.

Some of the more complex beam design problems, such as general biaxial bending and combined bending and torsion, are treated in Chapter 8. A brief treatment of

†The simple theory of bending will be reviewed briefly in Sec. 3.2.

plastic design will also be included in Chapter 8. In plastic design the required design load is multiplied by a load factor to give the required ultimate (plastic) strength, and the continuous beam or frame is chosen to have equal or greater ultimate load capacity. The *stress* due to bending is not calculated—in various sections of the beam it will be at or even slightly greater than the yield point at the ultimate strength of the structure. Plastic design is advantageous when fully continuous beams or frames are used. These are statically indeterminate in the elastic range, but the analysis problem becomes statically determinate when the ultimate strength is reached, another advantage for plastic design. However, allowable stress (elastic) design is customary and adequate for the design of statically determinate beams, such as were illustrated in Figs. 3.1 and 3.3(a) and (b). Elastic, or allowable-stress, design will be emphasized in this chapter, although a brief introduction to beam behavior in the inelastic range (Sec 3.3) will be included, because it is essential to an understanding of specification modifications of allowable stresses as well as to the study of plastic design.

3.2 ELASTIC BENDING OF STEEL BEAMS

The beginning design student should refresh his acquaintance with elementary beam theory as presented in texts on mechanics or strength of materials, and here only partially and briefly reviewed. Figure 3.6 shows a *unit* length of beam imagined as cut out of the complete beam at any location along the beam. It is acted upon by bending moment M and shear V, positive as indicated, and is shown in its undeflected straight position before loading and in its deflected and bent position after loading. Note that Y and y, positive as shown, are used to signify two different distances: deflection of the beam axis and distance within the beam cross section from the centroid, respectively. The *curvature* of the beam, or change in slope per unit length of beam, is denoted by ϕ, and the unit longitudinal *strain*, or change in length per unit length, of a horizontal beam fiber is therefore equal to

$$\epsilon = \phi y \qquad (3.1)$$

Since normal stress (f) is equal to the modulus of elasticity (E) multiplied by strain (ϵ), the stress due to bending is equal, by Eq. (3.1), to

$$f_b = E\phi y \qquad (3.2)$$

Thus the stress due to bending of a beam is known if the curvature is known. This fact could be of interest to a wire manufacturer who wished to know the diameter of a drum or reel on which drawn wire could be wound without inducing any permanent bend. Suppose, for example, that a wire with a diameter of 0.10 in. is wound on a reel having a diameter of 60 in. The curvature of the wire is equal to a unit length (1 in.) divided by the reel radius (30 in.). Thus

$$\phi = \tfrac{1}{30}$$

The maximum strain in the wire, by Eq. (3.1), is

$$\epsilon = \tfrac{1}{30}(0.05) = 0.00167$$

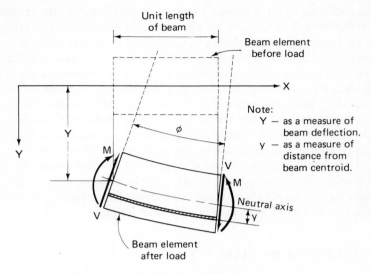

Fig. 3.6 Deformation of beam element.

The maximum stress, then, for steel wire, due to bending, for $E = 29,000$ ksi, is equal to

$$f_b = (29,000)(0.00167) = 48.3 \text{ ksi}$$

The stress of 48.3 ksi is greater than the yield point of carbon structural steel, but less than the elastic limit of most cold-drawn high-strength steel wires, for which the reel diameter would be satisfactory as it would not induce permanent bending deformation in the wire.

Equation (3.2) is convenient in the problem facing the wire manufacturer, but for the allowable stress design of steel beams the stress due to bending is usually calculated as a function of the bending moment, which is proportional to the curvature. The constant of proportionality between moment and curvature is EI, I being the moment of inertia of the cross section, as tabulated for all rolled sections in AISCM, pages 1-27 to 1-61. Thus the bending moment is

$$M = EI\phi \tag{3.3}$$

Combining Eqs. (3.3) and (3.2), the formula for stress in terms of bending moment is obtained:

$$f_b = \frac{My}{I} \tag{3.4}$$

Equation (3.4) is sometimes called the *beam equation*, and its use is restricted to the previously described simple bending theory. Initial beam selection is based on the maximum stress due to bending, for which $y = c$, where c is the maximum y distance from the centroidal (also neutral) axis of the beam to the extreme top or bottom fiber of the cross section. If the beam section is symmetrical about its x axis, c will be the same for the compression and tension extremities.

$$f_{b\,\text{max}} = \frac{M_{\text{max}}c}{I} \qquad (3.5)$$

To expedite the design selection of a beam for maximum bending moment, I and c are combined into a single parameter, the *section modulus*, denoted by S and equal to I/c.

Equation (3.5) then becomes simply

$$f_{b\,\text{max}} = \frac{M_{\text{max}}}{S} \qquad (3.6)$$

In allowable stress design the maximum stress due to bending (f_b) must be less than the allowable stress in bending (F_b) that is listed in the AISCS; thus the required section modulus in the design of a beam is

$$S_{\text{reqd}} = \frac{M_{\text{max}}}{F_b} \qquad (3.7)$$

Perhaps the most commonly used table in the AISCM is the one that tabulates the section modulus values for all rolled shapes used as beams. The tabulation excludes shapes normally used as columns, which are primarily those with flange widths approximately equal to the depth of the section.

In the AISCM read the first part of page 2-3, descriptive of the "Allowable Stress Design Selection Table." Look at the six-page tabulation of S_x (the section modulus for bending about the strong x axis of bending, pages 2-7 to 2-12). Skip the information on unbraced length, which will be discussed later.

Unless the beam is extremely short, it should be *chosen* for moment and *checked* for shear; that is, the shear stress f_v should be less than the allowable value F_v. The shear stress as computed by simple bending theory at any location in the beam web is given by

$$f_v = \frac{VQ}{It} \qquad (3.8)$$

where V = total resultant shear force on cross section
   Q = static moment, taken about the neutral axis, of that portion of the beam area beyond the point at which the shear stress is to be calculated
   t = web thickness where the stress is computed

The designer may use Eq. (3.8) for certain shear-dependent details, such as the welds connecting the web and flange of built-up beams, or the webs of unsymmetrical sections. But simpler expressions that approximate the web shear stress are specified for the usual beam design situation. For a W or C beam section, in simple bending, the shear stress is taken as the *average* value, with the web area taken as the web thickness times the full depth of the beam. In the design of built-up girders the area

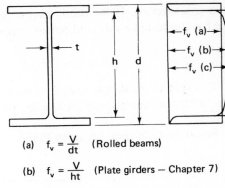

(a) $f_v = \dfrac{V}{dt}$ (Rolled beams)

(b) $f_v = \dfrac{V}{ht}$ (Plate girders — Chapter 7)

(c) $f_v = \dfrac{VQ}{It}$ (Shear dependent details or nonstandard shapes)

Fig. 3.7 Three alternatives for the estimate of web shear stress due to bending.

is based on the net depth of the girder between flange plates. Figure 3.7 illustrates the three alternatives just discussed.

3.3 INELASTIC BEHAVIOR OF STEEL BEAMS

If *static* working load is the primary basis for the choice of a beam, it may be selected on the basis of its maximum strength without recourse to stress calculations. This is standard procedure in *plastic design*, which is based not only on the plastic strength of the individual beam but on the ultimate strength of the complete continuous-frame structure. Moreover, the considerable variation in allowable stresses in *elastic design* can be justified, as will be shown, on the basis of maximum-strength criteria.

Referring back to Fig. 1.2, let it be assumed that the stress–strain diagram for structural steel consists simply of the two straight line portions out to the initiation of strain hardening and labeled "elastic range" and "plastic range," respectively. In this case, the maximum stress due to bending in a steel beam would not rise above the yield stress (F_y). Figure 3.8(b) shows the stress distribution assumed in allowable stress design. If the bending moment is increased above the amount that causes the stress to be equal to F_b, until the stress f_b just equals F_y, the stress diagram will continue to have the elastic, linear distribution shown in Fig. 3.8(b), and the bending moment will have reached M_y, the yield moment. Above the yield moment, the stress distribution will be as shown in Fig. 3.8(c), finally approaching in the limit the rectangular shape shown in Fig. 3.8(d), which is the plastic moment, M_p, the maximum attainable if no strain hardening were to occur.

The inelastic behavior of a steel beam is best illustrated by a plot of bending moment versus curvature. Above M_y the curvature is no longer linearly related to moment by the elastic relationship of Eq. (3.1). At M_y the curvature $\phi_y = F_y/Ec = M_y/EI$. Above M_y, at moments less than M_p, and using the notation shown in Fig. 3.8(c), the

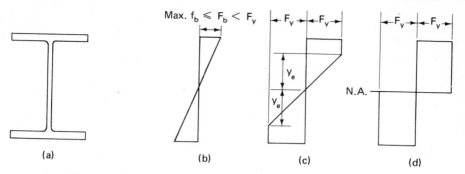

Fig. 3.8 Distribution of normal stress due to bending in elastic and inelastic ranges.

curvature is equal to F_y/Ey_e. M versus ϕ curves for circular, rectangular, and W beam shapes are shown in Fig. 3.9. In this plot each shape has the same section modulus and the same M_y. The ratio M_p/M_y is called the *shape factor*, which is a measure of the increase in plastic moment strength in comparison with the yield moment. The circular section is the most inefficient from an elastic point of view, although it absorbs more energy than the other shapes before reaching the yield moment. The average shape factor for the W shape is only 1.14, but this is a least-weight shape designed to be efficient in the elastic range of behavior. To ensure an approach toward M_p without local flange buckling or lateral buckling, the wide-flange shape must be "compact," as defined by AISCS, Sec. 1.5.1.4.1. If not compact, design is to be by the allowable-stress procedure, that is, based on M_y and not M_p.

Fig. 3.9 Inelastic M-ϕ curves for different cross sections.

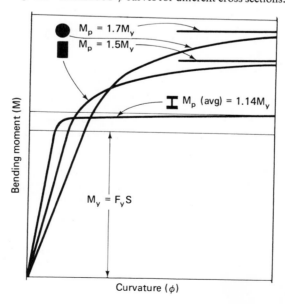

Just prior to the 1969 AISCS, the allowable-stress design procedure would have permitted bending moments tabulated below as $M_{allowable}$ for the four different shapes just discussed.

	$M_{allowable}$	M_{max}	$M_{max}/M_{allowable}$
Circular section	$0.9F_yS$	$M_p = 1.7F_yS$	1.89
Rectangular section	$0.75F_yS$	$M_p = 1.5F_yS$	2.00
Compact W shape	$0.66F_yS$	$M_p = 1.14F_yS$	1.73
Noncompact W shape	$0.60F_yS$	$M_y = F_yS$	1.67

The maximum moment is listed as the plastic moment for the circular, rectangular, and compact W shapes; but for the noncompact W shape there is no assurance of attaining and holding any moment greater than that for which the maximum stress first reaches the yield point. The strength load factors are tabulated in the third column. The load factors are in reasonable agreement, and illustrate how adjustments in allowable stress have been used to at least partially compensate for wide variations in the plastic ultimate strength. In the 1969 AISCS the allowable stress for the circular section has been reduced to $0.75F_y$ and the corresponding load factor has risen from 1.89 to 2.27.

To design a beam by the maximum-strength, or plastic-design, procedure, one first multiplies the required working loads by a design load factor, and determines the required maximum moment capacity at failure. This must be exceeded by the plastic moment (M_p) of the beam to be selected. As an index of M_p, the *plastic modulus, Z,* is introduced:

$$Z = \frac{M_p}{F_y} = S \times \text{(shape factor)} \tag{3.9}$$

The plastic modulus, Z, is tabulated for compact rolled W and S shapes on pages 2-14 to 2-20 of AISCM.

In summary, then, beam selection by plastic design involves

1. Determination of maximum moment at failure load, estimated as working load times a load factor.
2. Determination of required Z, which is moment determined in step 1 divided by yield stress F_y.
3. Selection of beam with greater Z than that required.
4. Checking the compactness of section and shear strength.

In the case of simple beams the selection will usually not be different from that obtained by allowable-stress design.

The real advantage of plastic design results from application to continuous beams or frames, concerning which a very brief treatment will be presented in Chapter 8.

3.4 ALLOWABLE STRESSES FOR ELASTIC DESIGN

In the usual allowable-stress design procedure, steel beams are selected so that the maximum normal and shear stress components due to bending do not exceed the AISCS allowable values which are specified in terms of the yield strength F_y. The allowable stresses in kips per square inch (ksi) are also tabulated for the various available yield points in the AISCS, Appendix A, pages 5-64 to 5-73. In some cases the listed values have been rounded off. For example, when the allowable stress is $0.66F_y$, it would be $0.66 \times 36 = 23.76$ ksi, but the allowable value as tabulated in the Appendix has been rounded off to 24 ksi. For convenience, the rounded-off values, as specified, will be repeated in appropriate locations in the following pages.

Compact Sections

AISCS, Sec. 1.5.1.4.1, page 5-17, states that the allowable stress on extreme fibers of *compact* hot-rolled sections (except A514 steel) symmetrical about and loaded in the plane of their minor axis shall be

$$F_b = 0.66F_y$$

F_y	36	42	45	50	55	60	65
F_b	24	28	29.7	33	36.3	39.6	42.9

The "basic" allowable stress in tension or compression is $0.6F_y$. The compact sections, for which a stress of $0.66F_y$ is permitted, are required to have relatively thicker flanges and webs than the noncompact and must have added safeguards against local and lateral buckling. The resulting bonus of 10 per cent in the allowable stress is for sections which, as discussed in Sec. 3.3, will develop their plastic moment strength, which is, on the average, 14 per cent greater than the yield moment strength. It is this difference between yield moment and plastic moment that establishes the logic of the stress increase.

The specific requirements are, by AISCS:

(a) The flanges shall be continuously connected to the web. (A built-up section with intermittent welds would not qualify.)

(b) For the unstiffened projecting elements of the compression flange, the width–thickness ratio shall be based on half the full flange width (AISCS, Sec. 1.9.2.1), and $b_f/2t_f$, therefore, shall not exceed $52.2/\sqrt{F_y}$, where b_f is the width of the compression flange, and t_f is the thickness of the compression flange.

F_y	36	42	45	50	55	60	65
$\dfrac{52.2}{\sqrt{F_y}}$	8.7	8.1	7.8	7.4	7.0	6.7	6.5

The "unstiffened" projecting element refers to the usual situation in standard hot-rolled sections, such as the W shape. Stiffened projecting elements include flanges with lipped edges, sometimes used in cold-formed shapes. The flange of a box beam flanked by two webs would also be considered a stiffened flange.

(c) For the stiffened projecting elements of the compression flange, the width–thickness ratio (b/t_f) shall not exceed

$$\frac{190}{\sqrt{F_y}} \qquad (b = \text{actual width of stiffened plate segment})$$

F_y	36	42	45	50	55	60	65
$\dfrac{190}{\sqrt{F_y}}$	31.7	29.3	28.3	26.9	25.6	24.5	23.6

(d) The depth–thickness ratio of the web shall not exceed

$$\frac{412}{\sqrt{F_y}}$$

F_y	36	42	45	50	55	60	65
$\dfrac{412}{\sqrt{F_y}}$	68.7	63.6	61.4	58.3	55.6	53.2	51.1

It should be noted that the foregoing rule for maximum d/t results from AISCS, Supplement No. 1 Eq. (1.5-4a) when $f_a = 0$. The case for which $f_a \neq 0$ will be considered in Chapter 5. The foregoing limitation on d/t is intended as a safeguard against premature web buckling in a beam or girder without transverse stiffeners.

(e) The compression flange shall be supported laterally, at intervals l_b, not to exceed either of the following limits:

$$\frac{l_b}{b_f} \leq \frac{76.0}{\sqrt{F_y}}, \qquad \frac{l_b d}{A_f} \leq \frac{20,000}{F_y}$$

F_y	36	42	45	50	55	60	65
$\dfrac{76.0}{\sqrt{F_y}}$	12.7	11.7	11.3	10.7	10.2	9.8	9.4
$\dfrac{20,000}{F_y}$	556	476	444	400	364	333	308

Transition Between Compact and Noncompact Members

If a beam with an unstiffened flange meets all the requirements for a compact section except that $b_f/2t_f$ exceeds $52.2/\sqrt{F_y}$, AISCS Sec. 1.5.1.4.2 provides for a transition in allowable stress between the values of $0.66F_y$ and $0.60F_y$. This does not apply to hybrid girders or to members of A514 steel, nor is the transition formula applied if it indicates an allowable stress less than $0.60F_y$, for which $b_f/2t_f$ would exceed $95.0/\sqrt{F_y}$. The transition formula [AISCS, Supplement No. 1 Eq. (1.5-5a)] is

$$F_b = F_y\left(0.733 - 0.0014\,\frac{b_f}{2t_f}\,\sqrt{F_y}\right)$$

A complete tabulation of the solution of this equation for the various yield stresses is provided in Appendix A of AISCS, pages 5-68 and 5-69.

Solid Round, Square, or Rectangular Sections

As discussed in Sec. 3.3, rectangular and round sections have plastic strength shape factors of 1.5 and about 1.7, respectively, with corresponding increases in ultimate moment capacity in relation to bending moment at initial yield. Thus, if designed by allowable-stress procedures, it is reasonable to permit an increase in allowable stress. W and S sections bent about the weak axis, provided the width–thickness ratio requirements for compact sections are met, are essentially similar in bending behavior to the solid rectangular section. Thus, excepting A514 steel, AISCS, Sec. 1.5.1.4.3, allows a stress in bending of $F_b = 0.75F_y$, provided, in the case of W and I sections, that the provisions of Sec. 1.5.1.4.1a and b are met:

F_y	36	42	45	50	55	60	65
F_b	27.0	31.5	33.8	37.5	41.3	45.0	48.8

When $b_f/2t_f$ is between $52.2/\sqrt{F_y}$ and $95.0/\sqrt{F_y}$, an otherwise compact shape, bent about the weak axis, may be designed for an allowable stress between $0.75F_y$ and $0.60F_y$, in accord with AISCS, Supplement No. 1 transition equation (1.5-5b):

$$F_b = F_y\left(0.933 - 0.0035\,\frac{b_f}{2t_f}\,\sqrt{F_y}\right)$$

Box Sections

Box sections are especially recommended for design situations involving incomplete lateral support. Failure by lateral-torsional buckling involves twisting in combination with lateral bending about the weak axis. Box sections are greatly superior to I or W (open) sections in both of these characteristics. Standard structural (box) tubes, as shown in Fig. 3.10(a), are catalogued in the AISCM, pages 1-104 to 1-106. Box beams also may be built up as a welded plate assemblage [Fig. 3.10(b)]. The AISCS would allow (1.5.1.4.1) the allowable stress for compact sections to be applied to box members, provided requirements (a) through (e) are met. However, requirement 1.5.1.4.1e is based on the properties of W shapes and is overly restrictive for box sections. For box sections with an overall depth no more than five times the

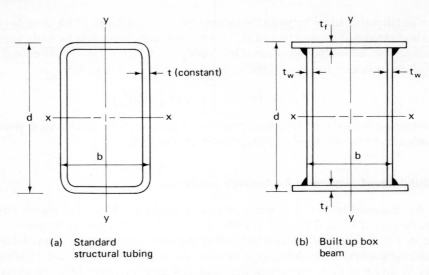

(a) Standard
structural tubing

(b) Built up box
beam

Fig. 3.10 Box beams.

transverse distance out to out of the webs, it can be shown that bracing to prevent lateral movement of the compression flange need be spaced no more than $650/F_y$ times the transverse distance out to out of the webs:

F_y	36	42	45	50	55	60	65
$\frac{650}{F_y}$	18.1	15.5	14.5	13.0	11.8	10.8	10.0

The foregoing is a recommendation of the authors and is not part of the AISCS†.
  Box sections that do *not* meet the requirements for compact sections (Sec. 1.5.1.4. 1), but do conform to the requirements of Sec. 1.9 ($b/t \leq 238/\sqrt{F_y}$), and which have lateral supports no farther apart than $2500/F_y$ times the transverse distance out to out of webs should be designed for the basic allowable stress of $0.6F_y$ (AISCS, Sec. 1.5.1.4.4).

F_y	36	42	45	50	55	60	65	100
$\frac{238}{\sqrt{F_y}}$	39.7	36.7	35.5	33.7	32.1	30.7	29.5	23.8
$\frac{2500}{F_y}$	69.4	59.5	55.6	50.0	45.5	41.7	38.5	25.0
$0.6F_y$	22.0	25.2	27.0	30.0	33.0	36.0	39.0	60.0

†As of 1972.

Miscellaneous Sections

In general, for any steel shape not explicitly cited, the maximum allowable tension in flexural members is $0.6F_y$, as per AISCS, Sec. 1.5.1.4.5. The accepted rounded-off values are listed in the last line of the previous paragraph.

W and C Shapes with Incomplete Lateral Support

W shapes are designed to be highly efficient when loaded in the plane of the web and supported laterally. When the lateral support is insufficient, the designer must decide as to which of several alternatives will afford the greatest economy. These include

1. Rearrange the framing to provide better lateral support.
2. Change to a box section, at higher cost per pound of material.
3. Select a W or C shape at an allowable stress reduced below the basic allowable of $0.6F_y$.

If the first choice is either not feasible or too costly, the third choice will probably prove the economical one if the reduction in allowable stress is relatively small. The stress-reduction alternative is covered by AISCS, Sec. 1.5.1.4.6, which provides two formulas, the first of which is in two parts, (1.5-6a) and (1.5-6b), The allowable stress is the *larger* obtained by trying the two formulas, but in no case is the allowable stress to be greater than the basic value of $0.6F_y$. In manual or slide-rule computation, the second formula is the easiest to compute; hence it is recommended that the designer check initially Formula (1.5-7), and if the result exceeds $0.6F_y$, there is no need to go to Formula (1.5-6). The AISCS Commentary should be referred to for justification for the use of two formulas.

When a beam fails by lateral-torsional buckling, it bends (buckles) about its *weak* axis, even though loaded normally in the strong plane so as to bend about its *strong* axis—which, indeed, it does, up to the critical load at which it buckles. When the beam buckles laterally about its weak axis, the loads also induce a torsional moment in the beam. The torsional resistance† of a W, S, or C section is made up of two parts: (1) the minimal torsion resistance that would be obtained under uniform torsion alone, plus (2) torsional resistance due to coupled bending of the flanges, inducing shears in the flanges that create a torsion couple. The torsion that accompanies lateral buckling is always nonuniform. Thus the resistance to lateral-torsional buckling of a W beam consists of three parts:

1. Lateral bending about the weak axis.
2. Uniform torsion resistance (St. Venant torsion).
3. Nonuniform torsion resistance (warping torsion).

†The torsional resistance of W shapes is explained in greater detail in Chapter 8.

In the interest of obtaining a simple procedure, AISCS Formulas (1.5-6) neglect the uniform torsion contribution, whereas Formula (1.5-7) neglects the nonuniform torsion contribution. Comparing the two contributions to torsional resistance, uniform torsion is relatively the smaller in thin-walled, deep sections, such as plate girders. Nonuniform, or warping torsion, is the smaller in thick-walled members of relatively long span. Section 1.5.1.4.6 is as follows:

1.5.1.4.6a Compression on extreme fibers of flexural members included under Sect. 1.5.1.4.5, and meeting the requirements of 1.9.1.2, having an axis of symmetry in, and loaded in, the plane of their web, and compression on extreme fibers of channels† bent about their major axis: the larger value computed by Formulas (1.5-6a) or (1.5-6b) and (1.5-7) as applicable (unless a higher value can be justified on the basis of a more precise analysis*), but not more than $0.60F_y$.

When
$$\sqrt{\frac{102 \times 10^3 C_b}{F_y}} \leq \frac{l}{r_T} \leq \sqrt{\frac{510 \times 10^3 C_b}{F_y}}$$
$$F_b = \left[\frac{2}{3} - \frac{F_y (l/r_T)^2}{1{,}530 \times 10^3 C_b}\right] F_y \qquad (1.5\text{-}6a)$$

When
$$\frac{l}{r_T} \geq \sqrt{\frac{510 \times 10^3 C_b}{F_y}}$$
$$F_b = \frac{170 \times 10^3 C_b}{(l/r_T)^2} \qquad (1.5\text{-}6b)$$

Or, when the compression flange is solid and approximately rectangular in cross-section and its area is not less than that of the tension flange
$$F_b = \frac{12 \times 10^3 C_b}{ld/A_f} \qquad (1.5\text{-}7)$$

In the foregoing,

l = distance between cross-sections braced against twist or lateral displacement of the compression flange

r_T = radius of gyration of a section comprising the compression flange plus one-third of the compression web area, taken about an axis in the plane of the web

A_f = area of the compression flange

$C_b = 1.75 + 1.05(M_1/M_2) + 0.3(M_1/M_2)^2$, but not more than 2.3,** where M_1 is the smaller and M_2 the larger bending moment at the ends of the unbraced length, taken about the strong axis of the member, and where M_1/M_2, the ratio of end moments, is positive when M_1 and M_2 have the same sign (reverse curvature bending) and negative when they are of opposite signs (single curvature bending). When the bending moment at any point within an unbraced length is larger than that at both ends of this length, the value of C_b shall be taken as unity. C_b shall also be taken as unity in computing the value of F_{bx} and F_{by} to be used in Formula (1.6-1a). See Sect. 1.10 for further limitation in plate girder flange stress.

†Only Formula (1.5-7) applicable to channels.
*See Commentary Sects. 1.5.1.4.5 and 1.5.1.4.6, last two paragraphs.
**C_b can be conservatively taken as unity. For smaller values see Appendix A, Fig. A1, p. 5-104.

For hybrid plate girders, F_y for Formulas (1.5-6a) and (1.5-6b) is the yield stress of the compression flange. Formula (1.5-7) shall not apply to hybrid girders.

1.5.1.4.6b Compression on extreme fibers of flexural members included under Sect. 1.5.1.4.5, and meeting the requirements of 1.9.1.2 but are not included in Sect. 1.5.1.4.6a:

$$F_b = 0.60F_y$$

provided that sections bent about their major axis are braced laterally in the region of compression stress at intervals not exceeding $76.0b_f/\sqrt{F_y}$.

AISCS Formulas (1.5-6) and (1.5-7) may be termed "semirational" in that they are simplifications of theoretical formulas for critical stress. The term C_b is intended to account for situations wherein the bending moment varies over the unbraced length. For uniform moment, $M_1 = -M_2$ and C_b is then unity, which is its minimum value. Values greater than unity should be restricted to situations for which the variation in bending moment within the unbraced length is linear, or nearly so, which will be the case if lateral bracing coincides with points of concentrated load application.

The seventh edition of the AISCM provides listings of both r_T and d/A_f for rolled members used as beams, thus facilitating the numerical application of Formulas (1.5-6) and (1.5-7).

AISCS, Sec. 1.5.1.4.6b, covers compression on extreme fibers of flexural members included under Sec. 1.5.1.4.5, but not included in Sec. 1.5.1.4.6a:

$$F_b = 0.60F_y$$

provided that sections bent about their major axis are braced laterally in the region of compression stress at intervals not exceeding $76.0b_f/\sqrt{F_y}$. This is a conservative requirement, since it was developed for the compact section.

Allowable Stress in Shear

Except in the case of very short spans, beams are usually selected on the basis of allowable stress in bending, then checked for the shear stress for which (Sec. 1.5.1.2) the allowable is $0.4F_y$. Reference is made to Sec. 1.10 for required reduction in this allowable value for very thin webs. The use of transverse stiffeners to increase the permissible shear stress in thin webs, also covered by Sec. 1.10, will be treated in Chapter 7. Formula (1.10-1), with $F_v = 0.4F_y$, requires that $C_v = 1.156$. As a/h becomes very large, the case with no transverse stiffeners, $k = 5.34$. These values of C_v and k then provide the following limit on h/t for which the full allowable shear stress $F_v = 0.4F_y$ may be used:

$$\frac{h}{t} \leq \frac{380}{\sqrt{F_y}}$$

Note that h is the clear depth between flanges, not the full depth d of the beam. For the various yield stresses:

F_y	36	42	45	50	55	60	65	90	100
$0.4F_y$	14.5	17.0	18.0	20.0	22.0	24.0	26.0	36.0	40.0
$\dfrac{380}{\sqrt{F_y}}$	63.3	58.6	56.6	53.7	51.2	49.0	47.1	40.0	38.0

The foregoing tabulation does not rule out the use of beam sections with greater h/t values, provided that a reduced allowable shear stress (less than $0.4F_v$) is specified according to the rules for thin web girder design. The reduced values are listed in AISCS, Appendix A, Tables 3-36 to 3-100, under the column captioned "a/h over 3."

3.5 LATERAL SUPPORT REQUIREMENTS

Most beams are designed by simple bending theory with the assumption of full lateral support and without any required reduction in allowable stress due to bending. Any beam with the compression flange attached securely to a floor or roof system that provides continuous or nearly continuous support meets these requirements.

Some conditions for which lateral support may be less than adequate include the following:

1. No positive connection between the beam and the load system that it supports, particularly if loads are vibratory or involve impact.

2. The lateral support system is of a removal type.

3. Lateral support system frames into a parallel system of two or more similarly loaded beams without positive anchorage. This possibility is illustrated in Fig. 3.11, which shows three alternative framing systems in plan view. In Fig. 3.11(a) the lateral support is inadequate for the reason cited. In Figs. 3.11(b) and (c) it is adequate, because of adjacent beams with wall anchorage or by virtue of K-bracing that limits motion, respectively.

Fig. 3.11 Plan view of beams with (a) inadequate and (b, c) adequate lateral support systems.

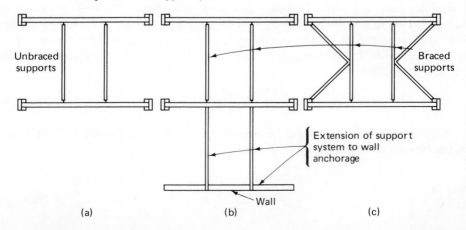

3.6 BEAM DEFLECTION LIMITATIONS

When plastered ceilings are attached to beams, the AISCS, Sec. 1.13.1, limits the maximum live load deflection to not more than $\frac{1}{360}$ of the span. This is an empirical relationship, based primarily on past experience, and intended to prevent or limit unsightly cracking in plastered surfaces.

Limits on deflection may be imposed indirectly by limitations on beam depth–span ratios. The interrelationship between span, stress, and deflection may be illustrated as follows. Consider the case of the uniformly loaded simple span; the deflection is

$$\Delta = \frac{5wl^4}{384EI} = \frac{5}{48}\frac{wl^2}{8}\frac{l^2}{EI} = \frac{5M_{max}l^2}{48EI} \tag{3.10}$$

The maximum stress due to bending is

$$f_b = \frac{M_{max}d}{2I}$$

which gives the following:

$$\frac{l}{d} = \frac{139,200}{f_b(l/\Delta)} \qquad \text{for } E = 29,000 \text{ ksi} \tag{3.11}$$

The AISCS Commentary (Sec. 1.13.1) suggests that the depth–span ratio be not less than $F_y/800$ for fully stressed beams and girders in floors, with the provision that this ratio may be decreased in the same ratio as the bending stress is decreased from that for a fully stressed beam. In the case of roof purlins, fully stressed, it is recommended that the depth–span ratio be not less than $F_y/1000$, except in the case of flat roofs, where special provisions to guard against "ponding" may be required.

For $f_b = F_b = 0.6F_y$, the foregoing recommendations of $F_y/800$ and $F_y/1000$ correspond to deflections of $l/290$ and $l/232$, respectively, for uniformly loaded simple beams. For the available yield points, these recommendations correspond to the following listed values of l/d:

F_y	36	42	45	50	55	60	65	90	100
$\dfrac{800}{F_y}$	22.2	19.0	17.8	16.0	14.5	13.3	12.3	8.9	8.0
$\dfrac{1000}{F_y}$	27.8	23.8	22.2	20.0	18.2	16.7	15.4	11.1	10.0

If water on a flat roof accumulates at a rate more rapid than it runs off, additional roof load develops due to the water that is "ponded" as a result of roof deflection, thus further aggravating the imbalance between rainfall and runoff rates. Failure may result. The problem is treated in the AISCS Sec. 1.13.3 and in the AISCS Commentary, pages 5-150 to 5-154.

In flat floor or roof systems of long span, the supporting girders should be cambered to minimize the ponding hazard if it exists, and to avoid any visually perceptible sag.

3.7 BEAMS UNDER REPEATED LOAD

Design of beams for repeated load by the allowable-stress-range procedure is essentially the same as for tension members, which has been covered in Chapter 2. Typical beam category designations are illustrated on page 5-112 in Appendix B of the AISCS.

In beam webs, in regions of relatively large shear stress combined with large normal stress due to bending, a simplification of procedure is provided in Appendix B of the AISCS on page 6-108. Instead of calculating the maximum principal tensile stress, the repeated load check is based simply on the maximum longitudinal *normal stress component* (Mc/I), but with a change of stress category if the shear stress at the same location exceeds $0.5F_v$.

In beam design it may be possible to effect economy by changing the arrangement of certain welded details so as to place the beam in the most favorable repeated load stress category.

3.8 BIAXIAL BENDING OF BEAMS

The W beam, or a modification thereof as shown in Fig. 3.12(b), is frequently used in design situations for which components of bending moment occur simultaneously about both the $x - x$ and $y - y$ axes. If the allowable stress about the two axes were the same, the design could be based simply on the maximum stress calculated by superposing the stresses caused by bending about each of the two principal axes:

$$f_b = \pm \frac{M_x}{S_x} \pm \frac{M_y}{S_y} \tag{3.12}$$

If the component of load in the weak plane is comparable in magnitude to that in the strong plane, a box section may be the preferred solution. If the lateral loads are relatively small, the W shape may be suitable. Because of the difference in allowable stresses about the two axes, an interaction formula—AISCS, Formula (1.6-2), with f_a set equal to zero —should be used. As modified, this becomes

$$\frac{f_{bx}}{F_{bx}} + \frac{f_{by}}{F_{by}} = 1.0 \tag{3.13}$$

Since F_{bx} and F_{by} depend on the selection of W shape, one cannot make a direct estimate of the required section modulus. However, an approximate approach may be made by introducing the following tentative assumptions:

$$F_{bx} = 0.60F_y, \qquad F_{by} = 0.75F_y, \qquad S_y = \frac{S_x}{C_n}$$

With these substitutions a formula for the preliminary selection on the basis of section modulus S_x is obtained:

$$S_x = \frac{M_x}{0.60F_y} + \frac{C_n M_y}{0.75F_y} \tag{3.14}$$

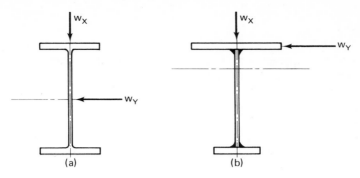

Fig. 3.12 Beams under biaxial load.

The coefficient C_n may be chosen on the basis of the following guidelines. If it is probable that the W section is in the 27- to 36-in. depth range, set $C_n = 7$, which is an approximation for the wider of the two W widths in this depth range. The wider sections will be most effective in resisting lateral loads and will also offer the greatest values of F_{bx}, which will be subject to reduction according to AISCS, Sec. 1.5.1.4.6. In the 16- to 24-in. depth range, there are four different sets of widths available and $C_n = 5$ could be used for preliminary trial selections. For depths of 14 in. and less, W column shapes are available as wide or wider than they are deep. For these, C_n ranges between 2.0 and 2.5 for most sections. Of course, if the lateral loads are only a very small proportion of the vertical loads, it will be more economical to use a section having C_n in the range 5 to 7.

If the lateral load is applied at or near the top flange, the use of an unsymmetrical section built up by welding three plates, as shown in Fig. 3.12(b), may be the most economical solution. The complex consideration of the torsion that is introduced may be avoided by calculating the stress in the top flange, due to lateral load, under the assumption that the top flange alone carries *all* the lateral load. This simplification of the analysis problem provides a safe design procedure. In applying the modified AISCS interaction formula (1.6-2), F_{bx} may be calculated by AISCS, Formula (1.5-6), with f_{bx} as determined for the compression flange. In the tension flange it may be assumed that the stress is unaffected by lateral load, but f_{bx} will be greater than in the compression flange.

Alternative approaches to the biaxial beam design problem are presented in Sec. 3.12.

3.9 LOAD AND SUPPORT DETAILS

Individual beams may carry concentrated loads and must be supported at or near their ends. Concentrated loads may be introduced directly into a beam web by means of a riveted, bolted, or welded beam web connection, and the loaded beam may in turn transmit its end reactions to columns or girders by means of similar connections. Beam web connections of the type just described are covered in Chapter 6.

When the end of a beam rests on masonry, it requires a bearing support, such as is shown in Fig. 3.13. Consideration must be given to the local compression stress in the web, just above the bearing block, and to the required thickness of bearing plate to spread the load to the masonry at a permissible pressure, f_p. The local concentration of compression in the beam web is assumed uniformly distributed over the distance $(N + k)$, as shown in Fig. 3.13. k is tabulated in the W shape detailing information tables in the AISCM, and is the distance from the flange face to the termination of the flange to web fillet. The compression stress along this line, equal to $R/(N + k)t_w$, must be kept below $0.75F_y$ (AISCS, Sec. 1.10.10). Note that the pressure is assumed to spread out from the edge of the bearing block at a 45° angle. If a support occurs away from the end of a beam, or if a local load is introduced at the top of a beam through a bearing block, the situation is similar, except the spread of load proceeds from each end of the bearing block and the compression stress is assumed to be $R/(N + 2k)t_w$. (Refer also to the AISCM, pages 2-22 and 2-23.)

Bearing plates must be thick enough to spread the reactions or concentrated loads to the masonry at allowable pressures, as specified in AISCS, Sec. 1.5.5. The required bearing plate thickness is determined by considering the bearing plate as a simple cantilever beam of length n, as shown in Fig. 3.13(a). The beam carries an upward load, assumed uniform, resulting from the masonry pressure. It is assumed that the plate will distribute the load satisfactorily if the stress due to bending in the cantilever beam is kept below $0.75F_y$ (AISCS, Sec. 1.5.1.4.3). A formula for bearing plate thickness is readily derived. Assume a unit width of cantilever beam:

$$M_{\max} = (f_p)(1)(n)\frac{n}{2} = \frac{f_p n^2}{2}$$

$$S = \frac{(1)(t^2)}{6}$$

$$f_{\max} = 0.75F_y = \frac{M}{S} = \frac{3f_p n^2}{t^2}$$

Solving for t, the required bearing plate thickness is

$$t_{\text{reqd}} = n\sqrt{\frac{4f_p}{F_y}} \tag{3.15}$$

For A36 steel this simplifies further to

$$t_{\text{reqd}} = \frac{n}{3}\sqrt{f_p} \tag{3.16}$$

where f_p is in kips per square inch. See, also, the AISCM, pages 2-82 and 2-83.

Fig. 3.13 Bearing supports at end of a beam.

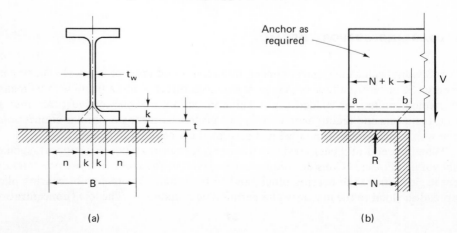

(a)   (b)

3.10 ALLOWABLE LOAD TABLES FOR BEAMS

The AISCM contains a wealth of design information that can expedite the design selection of simple beams. The tables provide for the direct selection in terms of uniform load and span of beams with yield stresses of either 36 or 50 ksi. Complete explanatory information is provided on pages 2-25 to 2-27 and the tables are on pages 2-28 to 2-81. The tables give the maximum unbraced length (L_c) in feet for compact shapes using $F_b = 0.66F_y$ and noncompact shapes where F_b is determined by AISCS Formula 1.5-5. Also given is the maximum unbraced length (L_u) in feet beyond which the allowable stress is less than $0.6F_y$. For unbraced lengths greater than L_u, graphs are provided (pages 2-87 to 2-105), which directly determine reduced allowable moments for allowable stresses less than $0.6F_y$.

If the load and/or support conditions are other than a simple beam under uniform load, the allowable load tables may be used in many instances by conversion of the different load situation to an "equivalent (uniform) tabular load." Conversion factors are given for uniformly spaced concentrated loads in the AISCM on page 2-26 and for a variety of other load and support conditions on pages 2-198 to 2-208.

Deflection coefficients for beams in terms of span and maximum stress due to bending on page 2-213 of the AISCM permit rapid calculation of center deflection of uniformly loaded beams. Modifying factors permit good approximations for other load conditions.

3.11 FLOW CHARTS FOR STEEL BEAM DESIGN

The flow charts presented on the following pages are based primarily on the AISCS and include revisions made during 1970 and 1971 as supplied by AISCS, Supplements No. 1 and No. 2. They also include, as noted, several recommendations for design by the authors.

The flow charts include:

Flow Chart 3.1: Allowable stress due to flexure, F_b. It will be called for during the use of both the steel beam and column flow charts, the latter in Chapters 4 and 5.

Flow Chart 3.2: Steel beam selection. This chart includes W, S, C, and box-shaped beams. Biaxial bending is included insofar as the section has at least one axis of symmetry. Insofar as channels are covered, lateral supports would need to be provided at support and load locations.

Flow Chart 3.1

(1) Allowable bending
stress about major
x-axis:

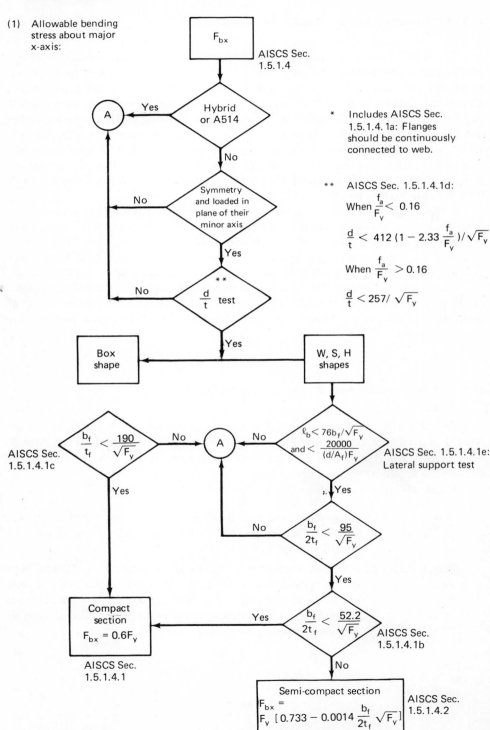

F_{bx}

AISCS Sec.
1.5.1.4

Hybrid
or A514 — Yes → Ⓐ

No

Symmetry
and loaded in
plane of their
minor axis — No

Yes

$\dfrac{d}{t}$ test ** — No

Yes

Box
shape ← W, S, H
shapes

* Includes AISCS Sec.
1.5.1.4. 1a: Flanges
should be continuously
connected to web.

** AISCS Sec. 1.5.1.4.1d:

When $\dfrac{f_a}{F_y} < 0.16$

$$\dfrac{d}{t} < 412\,(1 - 2.33\,\dfrac{f_a}{F_y})/\sqrt{F_y}$$

When $\dfrac{f_a}{F_y} > 0.16$

$$\dfrac{d}{t} < 257/\sqrt{F_y}$$

$\dfrac{b_f}{t_f} < \dfrac{190}{\sqrt{F_y}}$ — No → Ⓐ ← No

AISCS Sec.
1.5.1.4.1c

Yes

$\ell_b < 76 b_f/\sqrt{F_y}$
and $< \dfrac{20000}{(d/A_f)F_y}$

AISCS Sec. 1.5.1.4.1e:
Lateral support test

Yes

$\dfrac{b_f}{2t_f} < \dfrac{95}{\sqrt{F_y}}$ — No

Yes

Compact
section
$F_{bx} = 0.6F_y$ ← Yes — $\dfrac{b_f}{2t_f} < \dfrac{52.2}{\sqrt{F_y}}$

AISCS Sec.
1.5.1.4.1

AISCS Sec.
1.5.1.4.1b

No

Semi-compact section
$$F_{bx} = F_y\,[\,0.733 - 0.0014\,\dfrac{b_f}{2t_f}\,\sqrt{F_y}\,]$$

AISCS Sec.
1.5.1.4.2

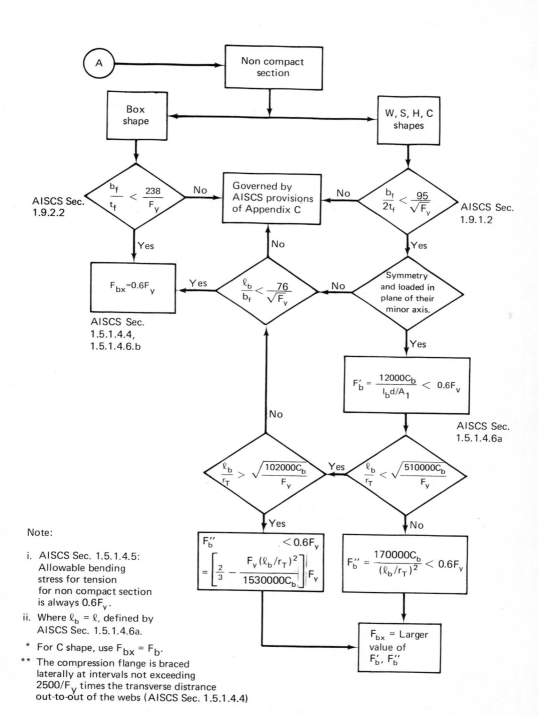

Note:

i. AISCS Sec. 1.5.1.4.5: Allowable bending stress for tension for non compact section is always $0.6F_y$.

ii. Where $\ell_b = \ell$, defined by AISCS Sec. 1.5.1.4.6a.

* For C shape, use $F_{bx} = F_b$.

** The compression flange is braced laterally at intervals not exceeding $2500/F_y$ times the transverse distrance out-to-out of the webs (AISCS Sec. 1.5.1.4.4)

Flow Chart 3.1 (continued)

(2) Allowable bending stress about minor y-axis:

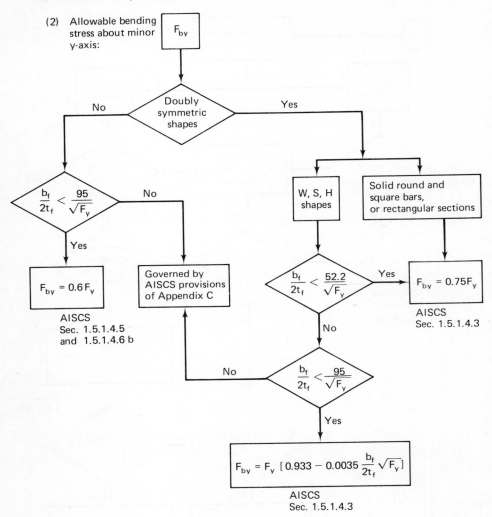

F_{by}

Doubly symmetric shapes

No — Yes

$\dfrac{b_f}{2t_f} < \dfrac{95}{\sqrt{F_y}}$

No

Yes

$F_{by} = 0.6F_y$

AISCS
Sec. 1.5.1.4.5
and 1.5.1.4.6 b

Governed by AISCS provisions of Appendix C

W, S, H shapes

Solid round and square bars, or rectangular sections

$\dfrac{b_f}{2t_f} < \dfrac{52.2}{\sqrt{F_y}}$

Yes

$F_{by} = 0.75F_y$

AISCS
Sec. 1.5.1.4.3

No

$\dfrac{b_f}{2t_f} < \dfrac{95}{\sqrt{F_y}}$

No

Yes

$$F_{by} = F_y \left[0.933 - 0.0035 \dfrac{b_f}{2t_f} \sqrt{F_y} \right]$$

AISCS
Sec. 1.5.1.4.3

Flow Chart 3.2

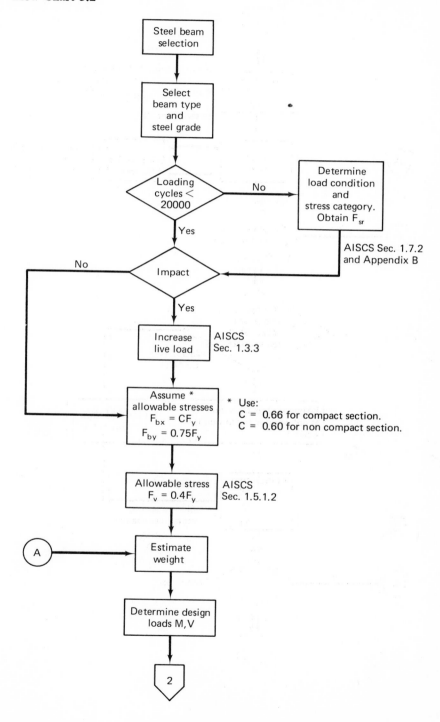

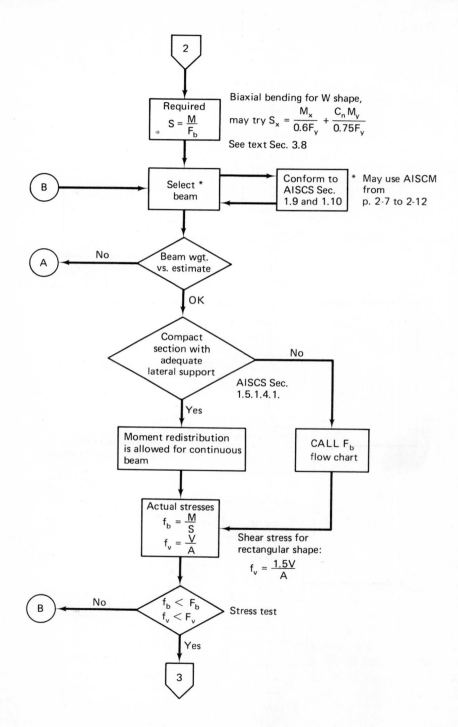

2

Required
$S = \dfrac{M}{F_b}$

Biaxial bending for W shape,
may try $S_x = \dfrac{M_x}{0.6F_y} + \dfrac{C_n M_y}{0.75F_y}$

See text Sec. 3.8

B

Select *
beam

Conform to
AISCS Sec.
1.9 and 1.10

* May use AISCM
from
p. 2-7 to 2-12

A ← No — Beam wgt.
vs. estimate

OK

Compact
section with
adequate
lateral support

No

AISCS Sec.
1.5.1.4.1.

Yes

Moment redistribution
is allowed for continuous
beam

CALL F_b
flow chart

Actual stresses
$f_b = \dfrac{M}{S}$
$f_v = \dfrac{V}{A}$

Shear stress for
rectangular shape:
$f_v = \dfrac{1.5V}{A}$

B ← No — $f_b < F_b$
$f_v < F_v$

Stress test

Yes

3

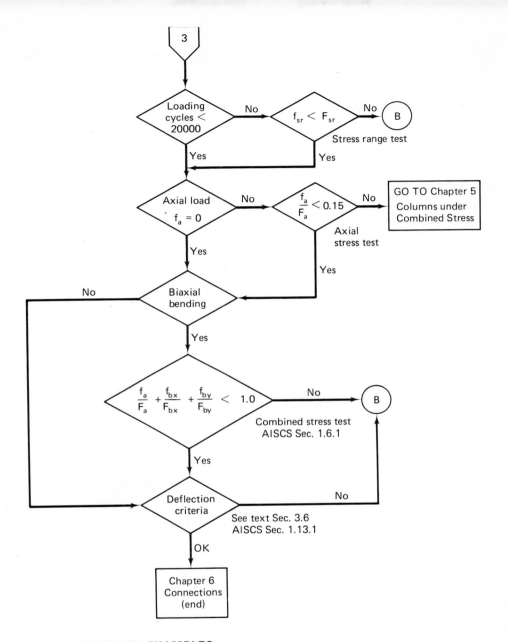

3.12 ILLUSTRATIVE EXAMPLES

The following illustrative problems demonstrate the use of the AISCS and AISCM in the selection and design of steel beams under some of the conditions that have been covered in this chapter. The problems are intended to do more than merely illustrate procedures; they also show how one may change conditions in some cases to achieve greater economy by altering certain details. The problems, in some cases, demonstrate the logic and judgment that may be applied to "zero in" on a solution when the first try turns out to be unsatisfactory. The reader is urged to make up hypothetical design situations on his own and carry out similar solutions.

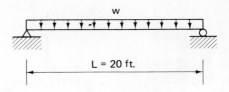

w

L = 20 ft.

Example 3.1

A simply supported beam (above) with a span of 20 ft is to carry a static uniform live load of 2.5 kips/ft. The flange is laterally supported by the floor system that it supports. Select the most economical W shape, using A36 steel.

Solution

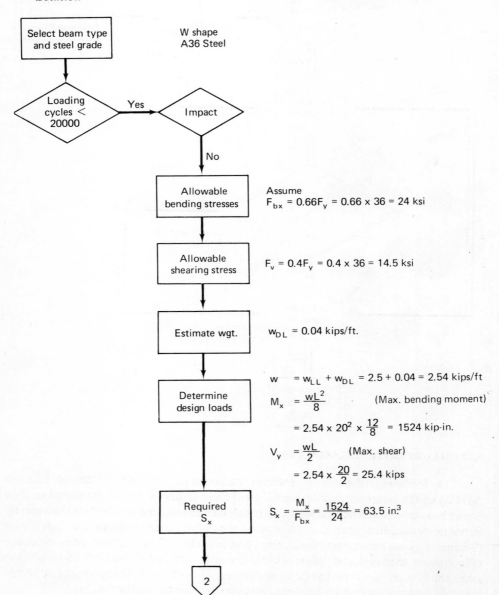

Select beam type and steel grade

W shape
A36 Steel

Loading cycles < 20000 — Yes → Impact

No

Allowable bending stresses

Assume
$F_{bx} = 0.66F_y = 0.66 \times 36 = 24$ ksi

Allowable shearing stress

$F_v = 0.4F_y = 0.4 \times 36 = 14.5$ ksi

Estimate wgt.

$w_{DL} = 0.04$ kips/ft.

Determine design loads

$w = w_{LL} + w_{DL} = 2.5 + 0.04 = 2.54$ kips/ft

$M_x = \dfrac{wL^2}{8}$ (Max. bending moment)

$= 2.54 \times 20^2 \times \dfrac{12}{8} = 1524$ kip-in.

$V_y = \dfrac{wL}{2}$ (Max. shear)

$= 2.54 \times \dfrac{20}{2} = 25.4$ kips

Required S_x

$S_x = \dfrac{M_x}{F_{bx}} = \dfrac{1524}{24} = 63.5$ in.^{3}

2

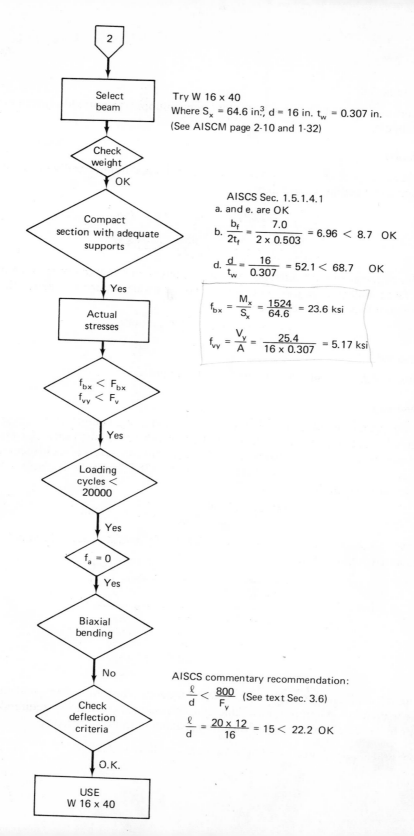

2

Select beam

Try W 16 x 40
Where S_x = 64.6 in.^{3}, d = 16 in. t_w = 0.307 in.
(See AISCM page 2-10 and 1-32)

Check weight

OK

Compact section with adequate supports

AISCS Sec. 1.5.1.4.1
a. and e. are OK

b. $\dfrac{b_f}{2t_f} = \dfrac{7.0}{2 \times 0.503} = 6.96 < 8.7$ OK

d. $\dfrac{d}{t_w} = \dfrac{16}{0.307} = 52.1 < 68.7$ OK

Yes

Actual stresses

$f_{bx} = \dfrac{M_x}{S_x} = \dfrac{1524}{64.6} = 23.6$ ksi

$f_{vy} = \dfrac{V_y}{A} = \dfrac{25.4}{16 \times 0.307} = 5.17$ ksi

$f_{bx} < F_{bx}$
$f_{vy} < F_v$

Yes

Loading cycles < 20000

Yes

$f_a = 0$

Yes

Biaxial bending

No

Check deflection criteria

AISCS commentary recommendation:

$\dfrac{\ell}{d} < \dfrac{800}{F_y}$ (See text Sec. 3.6)

$\dfrac{\ell}{d} = \dfrac{20 \times 12}{16} = 15 < 22.2$ OK

O.K.

USE
W 16 x 40

Example 3.2

Same as Ex. 3.1, but with intermittent lateral supports to provide equivalent full lateral support.

Solution

AISCS, Sec. 1.5.1.4.1e, provides criteria for the maximum laterally unsupported span segment of compression flange:

$$\frac{76b_f}{\sqrt{F_y}} = 12.7b_f{}^\dagger = 12.7 \times 7 = 89 \text{ in.}$$

$$\frac{20,000}{(d/A_f)F_y} = \frac{556\dagger}{d/A_f} = \frac{556}{4.54} = 122.5 \text{ in.}$$

Hence the maximum unsupported span segment of the compression flange is 89 in. The top flange is entirely in compression. Provide supports to divide the span in three equal-length segments. Then

$$l_b = \tfrac{240}{3} = 80 \text{ in.} < 89 \qquad \text{OK}$$

Example 3.3

Same as Ex. 3.1, but without any lateral support.

Solution

Note: A useful guideline that may be used in conjunction with the section modulus (S) requirement is the determination of the maximum value of d/A_f that would permit no reduction below $F_{bx} = 0.6F_y$, or, in the case of A36 steel, 22 ksi. Thus, for $F_b = 22$ ksi, $l_b = 240$ in., and $C_b = 1$ (AISCS, page 5-19), from AISCS, Sec. 1.5.1.4.6a, Formula (1.5-7),

$$\frac{d}{A_f} = \frac{12,000C_b}{F_b l_b} = \frac{12,000 \times 1}{22 \times 240} = 2.27 \qquad \text{(max)}$$

Referring to the Elastic Section Modulus Tables of AISCM, page 2-9, and to d/A_f as tabulated in AISCM Sec. 1

Shape	S_x	d/A_f
W 14 × 53	77.8	2.63
W 16 × 45	72.5	4.07
W 12 × 53	70.7	2.09
W 14 × 48	70.2	2.90

The above choices provide adequate section modulus. This tabulation indicates that a rather drastic reduction in allowable bending stress (F_b) would be required for all except the W 12 × 53 shape, for which no reduction at all is necessary.

†See AISCS, Appendix A, page 5-66. d/A_f values are listed in AISCM.

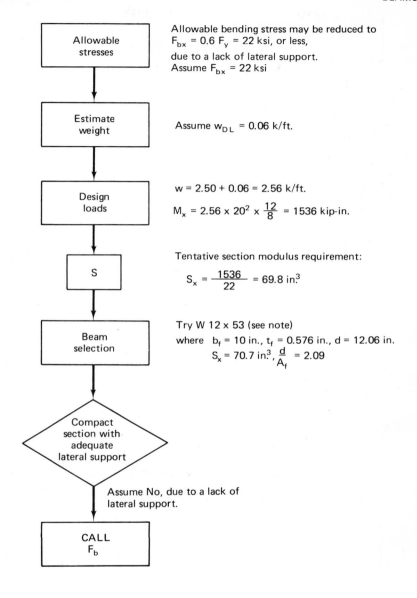

Allowable stresses

Allowable bending stress may be reduced to
$F_{bx} = 0.6\,F_y = 22$ ksi, or less,
due to a lack of lateral support.
Assume $F_{bx} = 22$ ksi

Estimate weight

Assume $w_{DL} = 0.06$ k/ft.

Design loads

$w = 2.50 + 0.06 = 2.56$ k/ft.

$M_x = 2.56 \times 20^2 \times \dfrac{12}{8} = 1536$ kip-in.

S

Tentative section modulus requirement:

$S_x = \dfrac{1536}{22} = 69.8$ in.^{3}

Beam selection

Try W 12 x 53 (see note)
where $b_f = 10$ in., $t_f = 0.576$ in., $d = 12.06$ in.
$S_x = 70.7$ in.^{3}, $\dfrac{d}{A_f} = 2.09$

Compact section with adequate lateral support

Assume No, due to a lack of lateral support.

CALL F_b

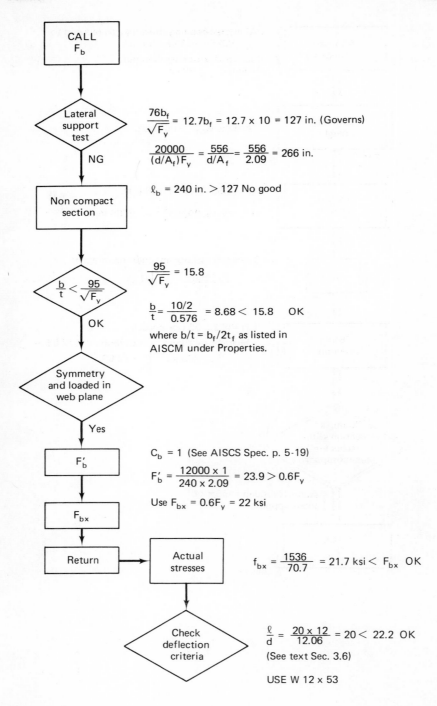

$$\frac{76b_f}{\sqrt{F_y}} = 12.7b_f = 12.7 \times 10 = 127 \text{ in. (Governs)}$$

$$\frac{20000}{(d/A_f)F_y} = \frac{556}{d/A_f} = \frac{556}{2.09} = 266 \text{ in.}$$

$\ell_b = 240 \text{ in.} > 127 \text{ No good}$

$$\frac{95}{\sqrt{F_y}} = 15.8$$

$$\frac{b}{t} = \frac{10/2}{0.576} = 8.68 < 15.8 \quad \text{OK}$$

where $b/t = b_f/2t_f$ as listed in AISCM under Properties.

$C_b = 1$ (See AISCS Spec. p. 5-19)

$$F_b' = \frac{12000 \times 1}{240 \times 2.09} = 23.9 > 0.6F_y$$

Use $F_{bx} = 0.6F_y = 22 \text{ ksi}$

$$f_{bx} = \frac{1536}{70.7} = 21.7 \text{ ksi} < F_{bx} \quad \text{OK}$$

$$\frac{\ell}{d} = \frac{20 \times 12}{12.06} = 20 < 22.2 \quad \text{OK}$$

(See text Sec. 3.6)

USE W 12 x 53

Example 3.4

Same beam selection as in Ex. 3.3, with uniform load replaced by an equivalent concentrated load at midspan that is repeated 3,000,000 times.

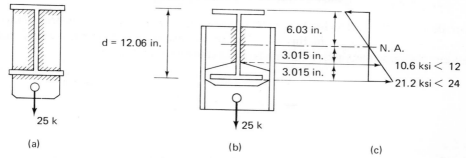

(a)   (b)   (c)

Solution

A midspan concentrated load of 25 kips causes the same moment as the uniform load of 2.5 kips/ft. The load is to be suspended with initial design of the loading detail in the sketch (a).

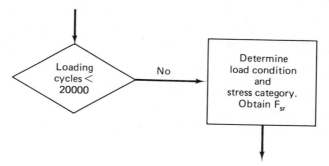

Loading condition is No. 4, as listed in AISCS, Appendix B, Table B1, for repetitions of load exceeding 2,000,000 times. Referring to Table B2, page 5-108, the situation is considered to be one of flexural stress adjacent to a welded transverse stiffener, for which the *stress category* depends on the magnitude of the shear stress:

$$f_v = \frac{12.5}{12.05 \times 0.345} = 3.0 \text{ ksi} < \frac{F_v}{2} = 7.25 \text{ ksi}$$

The *stress category* is now established as C, and from Table B3, Appendix B, then, the *allowable stress range* (F_{sr}) is found to be

$$F_{sr} = 12 \text{ ksi}$$

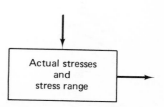

$$M_x = \frac{PL}{4} = \frac{25 \times 20 \times 12}{4} = 1500 \text{ kip-in.} \qquad \text{(max. live-load moment)}$$

$$f_{bx} = \frac{1500}{70.7} = 21.2 \text{ ksi} \qquad \text{(W } 12 \times 53)$$

(For design for repeated load, according to AISCS, Appendix B, the stress range $f_{sr} = 21.2 - 0 = 21.2$ ksi in this case. The *stress range* is the algebraic difference between the maximum and minimum stresses caused by live load.)

$f_{sr} = 21.2$ ksi $> F_{sr} = 12$ ksi   No good

Thus the design as it stands is unsatisfactory. Rather than select a heavier beam section, the problem can be solved by changing the weld detail as shown in the sketch (b) to reduce the live load stress level adjacent to the transverse weld to something less than F_{sr} (12 ksi), as shown in the sketch (c). Then the unwelded as-rolled beam surface at the location for maximum tension stress due to flexure has a stress category of A (Table B2, page 5-108) with an allowable stress range (f_{sr}) of 24 ksi. The shear stress should be rechecked on the basis of web material immediately adjacent to the weld.

$$f_v = \frac{12.5}{0.345 \times 6.03} = 6.01 \text{ ksi} < \frac{14.5}{2}$$

Hence, category C, Table B2, AISCS page 5-108 still holds and $F_{sr} = 12$ ksi. The beam design is OK. The design of the welded connection is deferred to Chapter 6.

Example 3.5

A simply supported beam with a span of 20 ft carries a uniform dead load of 1 kip/ft, including its own weight. A concentrated oblique load at midspan acts through the centroid, as shown. Determine economical beam selection for A36 steel if there are no intermediate lateral supports.

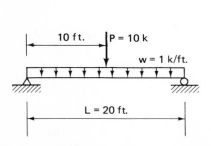

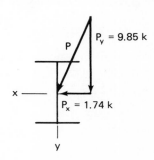

Solution

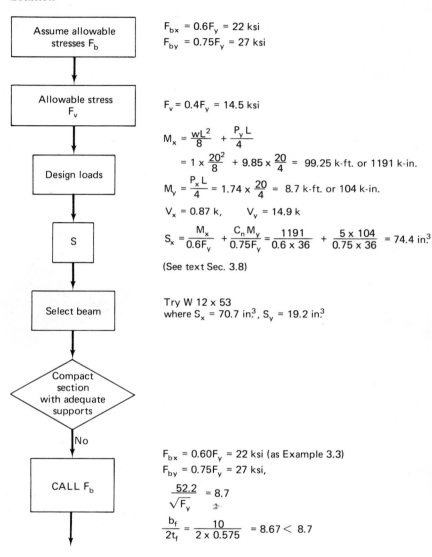

Assume allowable stresses F_b

$F_{bx} = 0.6F_y = 22$ ksi
$F_{by} = 0.75F_y = 27$ ksi

Allowable stress F_v

$F_v = 0.4F_y = 14.5$ ksi

Design loads

$M_x = \dfrac{wL^2}{8} + \dfrac{P_y L}{4}$

$ = 1 \times \dfrac{20^2}{8} + 9.85 \times \dfrac{20}{4} = 99.25$ k-ft. or 1191 k-in.

$M_y = \dfrac{P_x L}{4} = 1.74 \times \dfrac{20}{4} = 8.7$ k-ft. or 104 k-in.

$V_x = 0.87$ k, $V_y = 14.9$ k

S

$S_x = \dfrac{M_x}{0.6F_y} + \dfrac{C_n M_y}{0.75F_y} = \dfrac{1191}{0.6 \times 36} + \dfrac{5 \times 104}{0.75 \times 36} = 74.4$ in.^{3}

(See text Sec. 3.8)

Select beam

Try W 12 × 53
where $S_x = 70.7$ in.^{3}, $S_y = 19.2$ in.^{3}

Compact section with adequate supports

No

CALL F_b

$F_{bx} = 0.60F_y = 22$ ksi (as Example 3.3)
$F_{by} = 0.75F_y = 27$ ksi,

$\dfrac{52.2}{\sqrt{F_y}} = 8.7$

$\dfrac{b_f}{2t_f} = \dfrac{10}{2 \times 0.575} = 8.67 < 8.7$

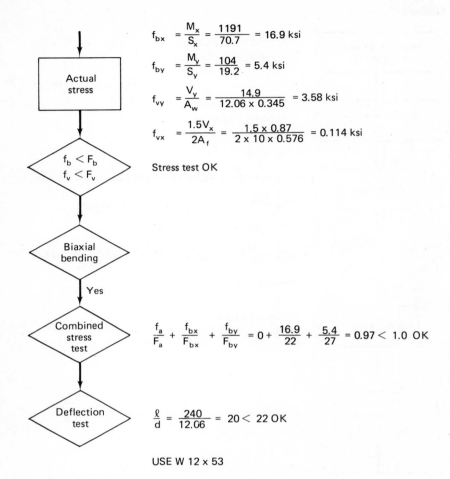

$$f_{bx} = \frac{M_x}{S_x} = \frac{1191}{70.7} = 16.9 \text{ ksi}$$

$$f_{by} = \frac{M_y}{S_y} = \frac{104}{19.2} = 5.4 \text{ ksi}$$

$$f_{vy} = \frac{V_y}{A_w} = \frac{14.9}{12.06 \times 0.345} = 3.58 \text{ ksi}$$

$$f_{vx} = \frac{1.5V_x}{2A_f} = \frac{1.5 \times 0.87}{2 \times 10 \times 0.576} = 0.114 \text{ ksi}$$

Stress test OK

$$\frac{f_a}{F_a} + \frac{f_{bx}}{F_{bx}} + \frac{f_{by}}{F_{by}} = 0 + \frac{16.9}{22} + \frac{5.4}{27} = 0.97 < 1.0 \text{ OK}$$

$$\frac{\ell}{d} = \frac{240}{12.06} = 20 < 22 \text{ OK}$$

USE W 12 x 53

Example 3.6

A simple span crane runway beam supports a moving load, transmitted by two wheels, as shown, of 80 kips, including impact. Using an unsymmetrical section of the type shown in Fig. 3.12(b), determine required plate sizes. Assume that the 16-kip lateral load is applied at the level of the top flange.

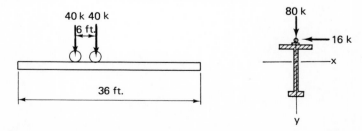

Solution

This problem will illustrate successive "cut-and-try" steps in approaching a final economical section. As discussed in Sec. 3.8, it will be assumed that all the lateral load is carried by the top flange to compensate for the fact that the torsional stresses will not be considered.

As a preliminary guide to determining a suitable size for the top (compression) flange, we make an initial "guess" that the total stress due to the sum of bending about the x and y axes will be 22 ksi and that in the compression flange about 75 per cent is due to the lateral load; that is, f_{bx} in compression is about 16.5 ksi.

As shown in many texts in structural analysis, the maximum moment due to a moving two-wheel load system is under the wheel nearest the center of the span when it and the center of gravity of the moving load system are equidistant from the center of the span, as shown in plan for the lateral load.

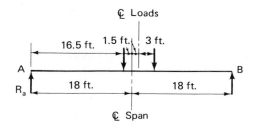

Design live loads:

$$R_{ax} = 16\frac{16.5}{36} = 7.33 \text{ kips}$$

$$R_{ay} = 80\frac{16.5}{36} = 36.65 \text{ kips}$$

$$M_x = 16.5R_{ay} = 604.73 \text{ kip-ft} = 7260 \text{ kip-in.}$$
$$M_y = 16.5R_{ax} = 121 \text{ kip-ft} = 1452 \text{ kip-in.}$$

Section modulus for top flange alone is

$$S_{yf} = \frac{b_f^2 t_f}{6}$$

$$f_{by} = \frac{M_y}{S_{yf}} = \frac{1452 \times 6}{b_f^2 t_f}, \qquad b_f = \sqrt{\frac{1452 \times 6}{f_{by} t_f}}$$

Try $t_f = 1.0$ in. and $f_{by} = 16.5$ ksi (assumed); then

$$b_f = \sqrt{\frac{1452 \times 6}{16.5 \times 1}} = 22.97 \text{ in.}$$

A. *First try*

Top flange:	*PL* 1 × 22
Bottom flange:	*PL* 1 × 8
Web:	*PL* $\frac{1}{2}$ × 34

Remarks:

    i. A trial overall depth of $\frac{1}{12}$ of the span, or $d = 36$ in., will be used.

    ii. $h/t = 68$, exceeding 63.3, as listed on "Allowable Stress in Shear" in Sec. 3.4, but shear stress will probably be well under 14.5 ksi.

    iii. If too narrow, bottom flange may buckle during erection, or vibrate during crane operation.

(1) Section properties:

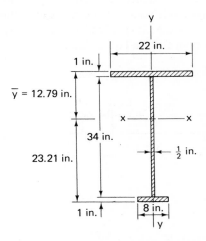

Locate neutral axis; take area moments about center of top flange.

Top flange	22.0×0	$= 0$
Bottom flange	8.0×35.0	$= 280.0$
Web	17.0×17.5	$= 297.5$
Total	47.0	577.5

$$\bar{y} = \frac{577.5}{47.0} + \frac{1}{2} = 12.79 \text{ in.} \quad \text{(see figure)}$$

Determine moment of inertia I_x and section moduli (neglect I_o of flanges):

Top flange	22.0×12.29^2	$= 3323$
Bottom flange	8.0×22.71^2	$= 4126$
Web	17.0×5.21^2	$= 461$
	$34.0^3 \times 0.5/12$	$= 1638$
Total	I_x	$= 9548$ in.^{4}

(*Note:* AISCM, pages 2-125 to 2-131, provides tables for more direct calculation of I_x.)

$$S_{xc} = \frac{9548}{12.79} = 746.5 \text{ in.}^3 \qquad \text{for compression flange}$$

$$S_{xt} = \frac{9548}{23.21} = 411.4 \text{ in.}^3 \qquad \text{for tension flange}$$

$$S_{yf} = \frac{22^2 \times 1.0}{6} = 80.7 \text{ in.}^3 \text{ for compression flange alone, about } y \text{ axis}$$

(2) Determine the allowable stress (F_{bx}) for the laterally unsupported top flange (see F_b flow chart):

$$F'_{bxc} = \frac{12,000C_b}{l(d/A_f)} = \frac{12,000 \times 1}{(36 \times 12)[36/(22 \times 1)]} = 17 \text{ ksi} < 22 \qquad \text{OK}$$

$$\text{top flange } I_y = \frac{22^3 \times 1}{12} = 887.3 \text{ in.}^4$$

$$\tfrac{1}{3} \text{ comp. area of web} = \frac{11.79 \times 0.5}{3} = 1.97 \text{ in.}^2$$

$$r_T = \sqrt{\frac{887.3}{23.97}} = 6.08 \text{ in.} \qquad \text{(AISCS, Sec. 1.5.1.4.6a)}$$

$$\frac{l}{r_T} = \frac{36 \times 12}{6.08} = 71.1$$

[Use AISCS, Formula (1.5-6a) with $C_b = 1$, since $53 < l/r_T < 119$.]

$$F''_{bxc} = 24.0 - \frac{71.1^2}{1181} = 19.72 \text{ ksi} < 22 \qquad \text{OK}$$

Use larger value of F'_{bxc} and F''_{bxc}; then $F_{bxc} = 19.72$ ksi.

(3) Determine stresses due to bending of the top flange (live load):

$$f_{bxc} = \frac{M_x}{S_{xc}} = \frac{7260}{746.5} = 9.7 \text{ ksi}$$

$$f_{byc} = \frac{M_y}{S_{yf}} = \frac{1452}{80.7} = 18.0 \text{ ksi}$$

At this point, since the total combined compressive stress in the top flange is 27.7 ksi, it is obvious that a new trial section will need to be chosen. However, as a guideline to the modification that may be needed, the stress due to tension will be calculated.

$$f_{bx} = \frac{M_x}{S_{xt}} = \frac{7260}{411.4} = 17.65 \text{ ksi}$$

B. *Choose second trial section*

Note that the neutral axis may be moved up and that the compression area and/or the depth of the section needs to be increased. Also F_{bxc} may as well be increased by widening the top flange. The stress due to dead load, previously omitted as a minor factor in preliminary calculations, will now be introduced. Omitting calculations that are a repetition in form of those made for the first trial, the following are found for the second trial section:

$$\begin{array}{ll} \text{Top flange:} & PL\ 1 \times 26 \\ \text{Bottom flange:} & PL\ 1 \times 7 \\ \text{Web:} & PL\ \tfrac{1}{2} \times 36 \end{array}$$

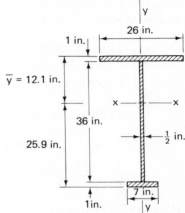

$$I_x = 10{,}995 \text{ in.}^4$$
$$S_{xc} = 908.7 \text{ in.}^3$$
$$S_{xt} = 424.5 \text{ in.}^3$$
$$S_{yf} = 112.7 \text{ in.}^3 \qquad \text{(top flange alone)}$$
$$r_T = 7.25 \text{ in.}$$
$$\frac{L}{r_T} = 59.6$$
$$F_{bxc} = 21.0 \text{ ksi}$$

Dead load moment (unit weight $= 51.0 \times 3.4 = 173$ lb/ft):

$$M_{DL} = \frac{0.173 \times 36^2 \times 12}{8} = 337 \text{ kip-in.}$$

(Note that while the maximum live load moment occurs 1.5 ft from the center of the span, at which point the dead load moment is slightly less than 337 kip-in., it will be a simple, conservative, and closely approximate procedure to add the two maximum values.)

Combining the maximum moments due to live and dead load,

$$M_x = 7260 + 337 = 7597 \text{ kip-in.}$$

The stresses due to bending are

$$f_{bxc} = \frac{M_x}{S_{xc}} = \frac{7597}{908.7} = 8.36 \text{ ksi}$$

$$f_{byc} = \frac{M_y}{S_{yf}} = \frac{1452}{112.7} = 12.88 \text{ ksi}$$

The combined compressive stress is now checked by the AISCS interaction formula (1.6-2), with $f_a = 0$ (AISCS, Sec. 1.6.1):

$$\frac{f_a}{F_a} + \frac{f_{bx}}{F_{bx}} + \frac{f_{by}}{F_{by}} = 0 + \frac{8.36}{21.0} + \frac{12.88}{22.0} = 0.98 < 1.0 \qquad OK$$

(Note that by AISCS, Sec. 1.5.1.4.3, assuming compliance with Sec. 1.5.1.4.1a and b, an allowable stress of 27 ksi could not in any case be used because the section is not doubly symmetric.)

Stress in tension flange due to bending:

$$f_{bxt} = \frac{M_x}{S_{xt}} = \frac{7597}{424.5} = 17.9 \text{ ksi} < 22 \text{ ksi} \qquad OK$$

Check shear stress. For this purpose, the moving wheel loads are placed at the extreme left end to produce the maximum shear, assumed equal to the maximum reaction:

$$V_y = \frac{80 \times 33}{36} + 0.173 \times 18 = 76.4 \text{ kips}$$

$$f_v = \frac{V_y}{A_w} = \frac{76.4}{18} = 4.24 \text{ ksi} < 14.5 \qquad OK$$

(The shear stress is so much below the allowable there is no point in checking the more accurate VQ/It formula.)

Check width–thickness ratios.

For flange:

$$\frac{b}{t} = \frac{13.0}{1.0} = 13$$

which is less than the 15.8 allowed for outstanding unstiffened elements (AISCS, Sec. 1.9.1.2, App. B, page 5-72). However, a 2 per cent overrun on specification requirements is usually considered tolerable.

For web:

$$\frac{h}{t} = \frac{36}{0.5} = 72$$

This exceeds the maximum value of 63.3, as tabulated in Sec. 3.4. However, for cases where the maximum shear stress is less than F_v for the case of no intermediate stiffeners (AISCS, Sec. 1.10.5.2, with $a/h = \infty$), it is permissible for h/t to exceed the values tabulated in Sec. 3.4. For a more complete discussion of this topic, reference should be made to Chapter 7, where treatment of the problem includes the use of intermediate stiffeners. At this time, for design of girders without intermediate stiffeners, use may be made of the extreme right column captioned "over 3" in the tables of AISCS, Appendix A, that starts on page 5-95. This column lists the maximum permissible shear stress for girders without intermediate stiffeners. In the present example, interpolating between the value of 13.0 for $h/t = 70$ and 11.4 for $h/t = 80$, the maximum allowable stress for $h/t = 72$ would be $F_v = 12.7$ ksi. Since this exceeds the maximum average shear stress of 4.24 ksi the design is satisfactory.

PROBLEMS

3.1. Assuming that full lateral support is provided, determine the maximum moment and shear in each of the following cases and make the most economical beam selection. Neglect beam dead weight in initial selection, then check stress due to both live and dead weight. Span is 22 ft in each case. Beams are simply supported at each end. Steel is ASTM A36 with $F_y = 36$ ksi.
  (a) Uniform load of 9 kips/ft.
  (b) Concentrated center load of 180 kips.
  (c) Three loads at $\frac{1}{4}$ points of 60 kips each.

3.2. A simply supported beam with a span of 36 ft carries a static uniform load of 3 kips/ft. The compression flange is laterally supported by the floor slab. Select the most economical W shape for a steel with $F_y = 42$ ksi. (Refer to Example 3.1.)

3.3. Same as Problem 3.2 but with intermittent lateral supports, as may be required, to provide the equivalent of full lateral support. (See Example 3.2.)

3.4. Same as Problem 3.2 but without any lateral support. (See Example 3.3.)

3.5. Design a simple beam for a span of 24 ft, using A36 steel, under a uniform load of 3 kips/ft and a concentrated load of 80 kips 2 ft from one end. Intermittent lateral supports are provided at 4-ft intervals.

3.6. Design a simple bearing block support for the heavily loaded end of the beam selected in Problem 3.5. Refer to Section 3.9 and to AISCM, pages 2-22 and 2-23, 2-82 and 2-83.

3.7. For the beam selected in Problem 3.5, compare (a) the maximum shear stress at the end of the beam nearest the concentrated load, using Eq. (3.8), (b) the average shear stress at the same location by AISCS, and (c) the average shear stress based on the clear depth web area, that is, the web area between the inner faces of the flanges.

3.8. Same as Problem 3.1 but with only a single lateral support at midspan.

3.9. Same as Problem 3.1, but with no lateral support between the ends of the beam.

3.10. Same as Problem 3.1(b), but with the 180-kip load repeated 1,200,000 times. Beam is fully supported laterally and it is assumed that the concentration of load at the center will require the use of a vertical stiffener similar to that sketched in Case 7, page 5-112 of AISCS. Use Example 3.4 as a guide. If the stress range permitted by the repeated load requirements controls the design, give special attention to a modification of stiffener details to improve the stress category as may be required.

3.11. A standard rectangular structural tube (see AISCM, pages 1-104 to 1-106) is used to span a highway with an effective simple span of 40 ft. In addition to its own dead weight, it supports a highway road direction sign having a total weight of 2400 lb, which is attached to the tube at its third points. The sign, measuring 4 ft in height by 20 ft in length, is centrally located and transmits to the tube a horizontal wind force that may be as great as 40 psf. This force is also introduced at the third points. Select a tube, using A36 steel. Do not neglect the wind force on the tube alone at each end of the span.

3.12. By use of Eq. (3.9) and the values of Section Moduli (S) and Plastic Moduli (Z) as provided in the AISCM, determine the "shape factor" for the following sections:

W   36 × 280
W   18 × 35
S   24 × 79.9
W   14 × 730

3.13. Using A36 steel, select the required size for the simply supported beam with overhang loaded as shown in the figure. Note that lateral support is provided only at the location of the 27 kip load and at the reaction supports.

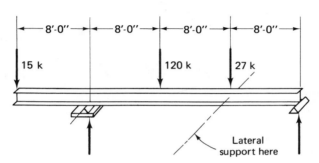

3.14. Select a W section, using A36 steel, for a span of 36 ft, to support a vertical load of 500 lb/ft in addition to its own dead weight. The beam, being exposed, must also resist a horizontal wind force of 30 psf.

3.15. A simple beam with a span of 20 ft carries a uniform vertical load of 1 kip/ft, including dead weight of beam, and a concentrated load at the center of 60 kips. Using A36 steel select:
(a) A W section for the vertical load alone.
(b) A W section for the vertical load plus a concentrated horizontal force of 6 kips, at the center of the span, and applied at the neutral axis of the beam.
(c) Same as (b), but with the horizontal force applied at the level of the top flange. (Assume that the top flange alone acts as a rectangular beam to resist the lateral force and keep the combined compressive stress within the allowable value of F_b.)

3.16. Same as Problem 3.15(c), but using a built-up welded member made of three plates similar to the section used in Example 3.6.

3.17. Similar to Example 3.6, except change span to 40 ft, total wheel load (without impact) 48 kips vertical and 12 kips horizontal. Apply the AISCS impact allowance for crane runway support girders to both the vertical and horizontal loads. The two wheel locations are 8 ft apart. Use A36 steel. Study Example 3.6 before attempting this problem, and try to use it as a guide in making a better initial selection of flange plate sizes.

4

COLUMNS
UNDER AXIAL LOAD

4.1 INTRODUCTION

Originally, the term "column" referred to a vertically upright compression member, such as are found in Egyptian, Greek, or Roman temples, built of hand-hewn segments of rock or marble. In today's usage, a column is not necessarily upright, and any compression member, horizontal, vertical, or inclined, is termed a column if the compression that it transmits is a primary factor affecting its structural behavior. If bending is also a major factor, the term *beam-column* may be used, and such members will be considered in Chapter 5.

For engineering design purposes, the axially loaded column is defined as one that transmits a compressive force whose resultant at each end is approximately coincident with the longitudinal centroidal axis of the member. Although there are no design loads that produce bending moment, there may be moments due to initial imperfections, accidental curvature, or unintentional end eccentricity. Such accidental bending moments unavoidably reduce the strength of the member, but are assumed to be taken care of in the design formula by an appropriate factor of safety.

The use of the column in structures has at times jumped ahead of design know-how. The tragic failure of the first attempt to build the Quebec Bridge, in 1907, has been attributed to faulty column design in which proportions and sizes were extrapolated beyond the range of previous experience. Unfortunately, the failure of compression members in structures is still too prevalent, and a real understanding of column behavior is most important to the structural engineer. Such an understanding provides a balance to the increasing complexity of analytical procedures and leads to an intelligent use of design specifications.

The study of the column introduces the phenomenon of *buckling*, during which a member experiences deflections of a totally different character than those associated with the initial loading. Thus, when an axially loaded column is first loaded, it simply shortens or compresses in the direction of the load. If and when a buckling load is reached, the shortening deformation stops, and a sudden lateral and/or twisting deformation occurs in a direction normal to the column axis. These added deflections limit the column's ability to carry added load.

The strength of a tension member is independent of its length, whereas for the column both the strength and the mode of failure are markedly dependent on length. A very short and compact column built of any one of the common metals will develop about the same strength in compression as it will in tension. But if the column is long, it will fail at a load that is proportional to the bending rigidity of the member, EI, and independent of the strength of the material. Thus a very slender column of steel with a yield stress of 100 ksi has no more column strength than one with a yield stress of 36 ksi. In each case the strength is determined by the Euler column formula that was developed more than 200 years ago.

4.2 BASIC COLUMN STRENGTH

The buckling strength of a column decreases with increasing length. Beyond a certain length the buckling stress falls below the proportional limit of the material stress–strain curve, and buckling for any longer column is termed elastic. For such a slender column the buckling load is given by the Euler formula,

$$P_e = \frac{\pi^2 EI}{l^2} \tag{4.1}$$

Again note that the yield stress, F_y, does not appear in Eq. (4.1). It plays no part in determining the strength of a very long column. Thus a slender aluminum alloy column, according to Eq. (4.1), will buckle at about one third of the load of its steel twin counterpart—not because of any weakness in the material, but simply because the elastic modulus E of aluminum alloy is about one third that of steel. The aluminum column would also weigh about one third that of steel, and by reshaping the design of the cross section the strength can be increased, up to a point, with *no increase in the weight of the member*. The limit of such increase occurs when the material gets so spread out, and correspondingly thin, that *local buckling* occurs prior to general buckling of the entire member.

The Euler load, P_e, is a load that will just hold the column in the deflected shape shown in Fig. 4.1. At every point along the column the external applied moment Py is equal to the internal resisting moment, $EI\phi$, where ϕ is the column curvature at the corresponding point. (Refer to Section 3.2.)

Fig. 4.1 Buckled shape of a hinged end column.

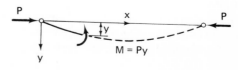

If both sides of Eq. (4.1) are divided by A and the relationship $I = Ar^2$ is introduced, r being the radius of gyration of the cross section, the buckling load is expressed in terms of the buckling stress, F_e:

$$F_e = \frac{P_e}{A} = \frac{\pi^2 EI}{Al^2} = \frac{\pi^2 E}{(l/r)^2} \tag{4.2}$$

Equation (4.2) can be modified so as to apply to other end conditions, such as free or fixed, by use of the effective length factor K. For purely flexural buckling, Kl is the length between inflection points and is known as the effective length. Thus Eq. (4.2) becomes

$$F_e = \frac{\pi^2 E}{(Kl/r)^2} \tag{4.3}$$

For example, if a column is fixed against rotation and translation (lateral movement) at each end, it will buckle with points of inflection at the quarter points, as shown in Fig. 4.2(a), and the effective length factor is 0.5. Consequently, according to Eq. (4.3), a very slender column with fixed ends that buckles elastically will be four times stronger than the same column with hinged ends. But if one end is fixed and the other free both with respect to rotation and translation, as shown in Fig. 4.2(b),

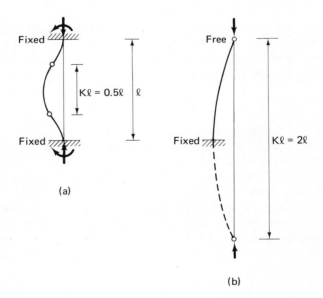

Fig. 4.2   Example illustrating the effective length concept.

there is an imaginary point of inflection at a distance l below the column base, and the effective length factor is 2.0. Such a column has only one fourth the strength of the same member with hinged ends. Table 4.1 shows these and other cases and lists recommended modified values for design usage.

l/r is termed the slenderness ratio and is almost universally used as a parameter in

Table 4.1

*Effective length factors K for centrally loaded columns
with various idealized end conditions.*

	(a)	(b)	(c)	(d)	(e)	(f)
Buckled shape of column is shown by dashed line.						
Theoretical K value	0.5	0.7	1.0	1.0	2.0	2.0
Recommended K value when ideal conditions are approximated	0.65	0.80	1.2	1.0	2.10	2.0
End condition code	Rotation fixed Translation fixed	Rotation free Translation fixed	Rotation fixed Translation free	Rotation free Translation free		

terms of which the column-strength curve may be drawn graphically or expressed analytically by a column-strength formula. Figure 4.3 shows typical column-strength curves for steel. The strengths of the very short and very long column are expressed by F_y and F_e, respectively. In the intermediate range, the transition from F_y to F_e depends on a complex mix of factors—initial curvature, accidental end eccentricity, and residual stress—and is usually expressed empirically by means of parabolic, straight-line, or more complex expressions.

In the case of structural steel, the presence of residual stress in rolled shapes has been shown to be the dominant factor influencing the shape of the transition curve between very short and very long columns.† Residual stresses are locked in a member as a result of uneven cooling after rolling, welding, oxygen cutting, or by cold-straightening operations.

On the basis of column tests, as well as measurements of residual stresses in rolled shapes, the Column Research Council (CRC) proposed in 1960 an empirical transition for the short and intermediate range, from F_y for $Kl/r = 0$, to the Euler curve (F_e) at $F_y/2$ as given by Eq. (4.4). For $Kl/r < C_c$,

$$F_c = \left[1 - \frac{1}{2C_c^2}\left(\frac{Kl}{r}\right)^2 \right] F_y \qquad (4.4)$$

†For a more complete treatment of inelastic buckling and the residual stress effect, refer to the *Column Research Council Guide to Design Criteria for Metal Compression Members* [1.6].

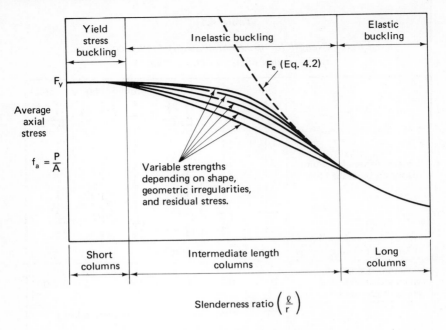

Fig. 4.3 The column strength curve.

where F_c = column strength, expressed in kips per square inch

$C_c = \sqrt{2\pi^2 E/F_y}$

For $Kl/r \geq C_c$, the Euler formula, Eq. (4.3), should be applied.

In the selection of columns to support design loads, the allowable stress, F_a, is determined by dividing Eq. (4.4) or (4.3) by an appropriate safety factor. This will be discussed in Sec. 4.7 and will be followed by various illustrative design examples.

4.3 EFFECTIVE LENGTH OF COLUMNS

The basic concept of effective length has been explained in the previous section, and certain special cases were shown in Table 4.1. A more general evaluation of effective length factors for columns in continuous frames is available through the use of the alignment charts shown in Fig. 4.4, as originally presented in the CRC *Guide*. One of these is also in the AISCS Commentary, Fig. C-1.8.2, page 5-139, for the case with sidesway permitted, as in Fig. 4.4(b).

These charts are functions of the I/L values of adjacent girders (beams), which are assumed as rigidly attached to the columns. A conservative assumption is made that all columns in the portion of the framework under consideration reach their individual buckling loads simultaneously. The charts are based upon a slope-deflection analysis that includes the effect of column load. In Fig. 4.4(a) and (b) the subscripts

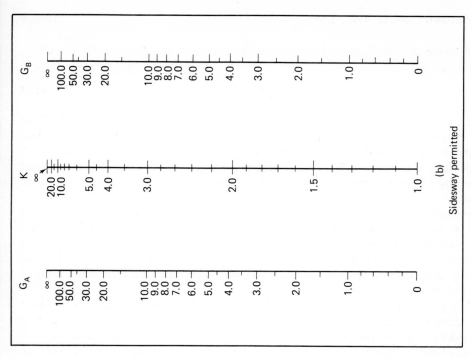

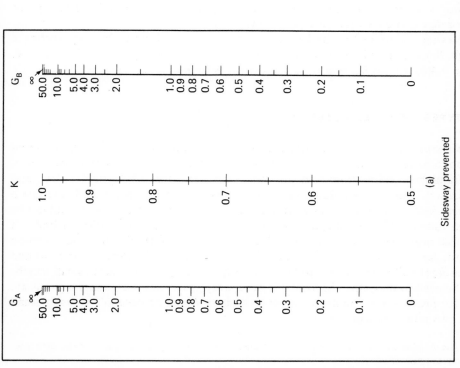

Fig. 4.4 Charts for effective length of columns in continuous frames.. (From the Column Research Council Guide, therein by courtesy of Jackson and Moreland Division of United Engineers and Constructors, Inc.)

A and *B* refer to the joints at the two ends of the column that is under construction. *G* is defined as

$$G = \frac{\sum I_c/L_c}{\sum I_g/L_g} \tag{4.5}$$

In Eq. (4.5) the $\sum$ indicates a summation of all members rigidly connected to that joint (*A or B*) and lying in the plane in which buckling of the column is being considered. I_c is the moment of inertia and L_c is the unsupported length of a column section. I_g is the moment of inertia and L_g is the unsupported length of a girder (beam) or other restraining member. I_c and I_g are taken about the axis perpendicular to the plane of buckling.

In connection with the use of these charts the following recommendations are made by the CRC *Guide* and are referred to in the AISCS Commentary,† Sec. 1.8. For a column base connected to a footing or foundation by a frictionless hinge, *G* is theoretically infinite, but should be taken as 10 in design practice. If the column base is rigidly attached to a properly designed footing, *G* approaches a theoretical value of zero, but should be taken as 1.0.

For greater accuracy, the girder stiffness I_g/L_g of Eq. (4.5) should be multiplied by a factor when certain conditions at the far end are known to exist. For the cases with sidesway prevented [Fig. 4.4(a)],† the appropriate multiplying factors are 1.5 for end of girder (beam) hinged, and 2.0 for far end of girder fixed. For the case with sidesway not prevented [Fig. 4.4(b)],† the multiplying factors are 0.5 for far end of girder hinged, and 0.67 for far end of girder fixed.

Having determined G_A and G_B for a column, *K* is obtained by constructing a straight line between the appropriate points on the scales for G_A and G_B. For example, in Fig. 4.4(a) if G_A is 0.5 and G_B is 1.0, then *K* is found to be 0.73.

4.4 TYPES OF STEEL COLUMNS

Various cross-sectional column shapes are shown in Fig. 4.5. The column cross section to be used will be conditioned by the magnitude of the load and by the type of end framing or connections that are most convenient for the particular structural application. Thus a pipe column with top and bottom bearing plates and based on a concrete footing is excellent as an isolated unit supporting a beam, but would not be as well suited to a truss with gusset plate connections. In general, within the limits of available clearance and with an eye to thickness limitations, the designer uses a section with the largest possible radius of gyration, thus reducing the slenderness ratio and increasing the allowable stress. In building design, of course, with rather small lengths and the need to maximize the available occupancy space, heavy compact sections are used for heavy loads. Similarly, in exposed areas it may be desirable to use a small section to minimize wind loads.

---

†The AISCS Commentary, page 5-139, includes only the alignment chart for the case with sidesway not prevented.

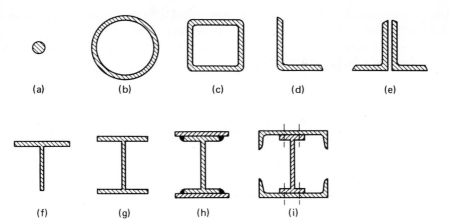

(a)   (b)   (c)   (d)   (e)

(f)   (g)   (h)   (i)

Fig. 4.5   Types of steel columns.

(a) *Round, Solid Bars*

The radius of gyration (r) for a solid round cross section is equal to

$$r = \frac{d}{4} = \frac{A}{\pi d} \tag{4.6}$$

where d = diameter of the solid round bar
                     A = cross-sectional area of the bar

In other less compact column cross sections that do not have a fixed geometry, the area can be varied independently of the radius of gyration, offering more of an advantage in connection with the substitution of higher-strength steels.

The solid round bars of high-strength steel have found particular use as main column elements of tall TV or radio towers, which have been built to heights of more than 2000 ft. Aside from its increased strength, high-strength steel reduces dead weight, which is particularly beneficial when seismic forces are being considered. Similarly, the use of a round section of relatively small diameter minimizes wind force and reduces added weight due to ice formation. Obviously, lateral force is a problem in the design of the individual column element acting as a beam, as in an exposed tower structure. The weight and area of round and square solid bars are tabulated in the AISCM, pages 1-110 and 1-111.

(b) *Steel Pipes*

The steel pipe as shown in Fig. 4.5(b) is more efficient than the solid section, since the radius of gyration may be increased almost independently of the cross-sectional area, thus reducing l/r and increasing the allowable stress, F_a. The possibility of local buckling must be considered if the wall thickness in comparison to the pipe diameter becomes overly small. According to American Iron and Steel Institute

(AISI) specifications, which deal with thin-walled and cold-formed steel members, the full allowable column stress may be used as long as the D/t ratio is less than $3300/F_y$, as tabulated below for the commonly used yield stresses. (D is the mean diameter of the pipe and t is the wall thickness.)

Yield stress (ksi)	36	42	45	50	55	60	65	90	100
Limit on $\dfrac{D}{t}$ $\left(\dfrac{3300}{F_y}\right)$	92	79	73	66	60	55	51	37	33

Material cost per pound for tubular sections almost always exceeds that of standard rolled sections, and the end connections in a framed structure will require special consideration. Advantages are the same as those listed for solid round bars. If the ends are hermetically sealed to prevent access of air, the pipe interiors need not be treated to prevent corrosion. Reference should be made to the AISCM, pages 1-101 and 1-102, for dimensions and properties of selected available standard sizes of steel pipe, ranging from the smallest up to 12 in. in diameter. Such pipe are available with a yield stress of 36 ksi, and all have D/t ratios well within the tabulated limits of $3300/F_y$, even for the high-strength steels. The AISCM provides direct determination of allowable column loads for steel pipe with a yield stress $F_y = 36$ ksi. (See pages 3-37 to 3-40.)

(c) Box Sections and Structural Tubes

The box section shown in Fig. 4.5(c) is of the standard types available, square up to 10×10 in. and rectangular up to 12×8 in., with properties of cross section as listed in the AISCM, pages 1-103 to 1-106. Larger sizes may be built up by welding various combinations of plates, angles, or channels. As a compression member, the square tube combines the effectiveness of the hollow steel pipe with the advantage of simpler end connection details in usual building-frame applications. Direct determination of allowable column loads is available for standard square and rectangular structural tubes with yield stresses of either 36 or 46 ksi. (See AISCM, pages 3-41 to 3-56.)

(d) Angle Struts

Single-angle struts, as shown in Fig. 4.5(d), are satisfactory as secondary members for light loads. If they can be used in a situation where the load can be brought uniformly into both angle legs at each end or where the ends can be prevented from rotation or twist by a rigid end connection attachment to a heavy member or footing, they may be designed by usual procedures for centrally loaded columns, provided width–thickness ratio limitations are met.

Double-angle struts, as shown in Fig. 4.5(e), are often used in single plane trusses. Frequent "stitching" must be provided by means of bolts or rivets to ensure that the two angles act as a single unit. (See AISCS, Sec. 1.18.2.6, page 5-48.) End connections should be designed so as to result in uniform distribution of load. The cross-sectional properties of selected double-angle sections are tabulated in the AISCM, pages 1-64 to 1-79, for members made up either of two unequal leg angles or two equal leg angles. Direct determination of allowable column loads is made possible by reference to the AISCM, pages 3-60 to 3-93, for double-angle struts of either 36 or 50 ksi yield stress material. No tabulation of loads is given for single-angle struts, and the AISCM (page 3-59) puts in a word of caution regarding such members: "it is virtually impossible to load such struts concentrically." Nevertheless, the single-angle strut will achieve equivalence of concentric loading if its ends can be restrained from rotation or twist about any axis.

The AISCS, Sec. 1.9.1.2, page 5-25, provides that the single-angle struts as well as double-angle struts with separators subject to axial compression shall be considered as fully effective when the ratio of width to thickness is not greater than $76/\sqrt{F_y}$ (for struts comprising double angles in contact the ratio shall be not greater than $95/\sqrt{F_y}$); otherwise, the allowable compression stress shall be modified by the appropriate reduction factor as provided in the AISCS, Secs. C2 and C5, pages 5-115 and 5-117. The foregoing width–thickness limits are tabulated in Appendix A of the AISCS on page 5-72.

(e) Structural Tees

Structural tees, as shown in Fig. 4.5(f), are often used as chord sections in light welded trusses, with double-angle struts welded to the tee web. Structural tees are made by splitting W shapes, or standard beams, longitudinally; hence they have stems (webs) considerably thinner that the flange. The restraint of the flange with respect to buckling of the stem permits the use of greater width–thickness ratios for the stem of the tee in comparison with the angle struts, as also provided by the AISCS, Sec. 1.9.1.2, permitting a width–thickness ratio in the stem up to $127/\sqrt{F_y}$, as tabulated on page 5-72, Appendix A, of the AISCS. For ratios greater than this limit, the allowable compression stress is modified in accordance with Secs. C2 and C5 of the AISCS, pages 5-115 and 5-117.

(f) Wide-Flange Shapes

Wide-flange W, M, or HP shapes are doubly symmetric, as shown in Fig. 4.5(g), and are rolled in a wide range of weights and sizes. These are therefore suited to a correspondingly great range of column loads and lengths and find frequent use in building construction. The AISCM provides tables for the direct determination of allowable axial loads for W shapes (pages 3-11 to 3-35).

The W 14 series (see pages 1-34 to 1-37 of the AISCM) provides a wide range of coverage of sections (some with exceptionally wide flanges to balance r about the two axes), which are especially suited to column requirements for tall multistory building frames.

In areas remote from mills that produce heavy W sections, equivalent sections are produced by means of continuous longitudinal welds joining three plate segments and designated WW. Combination sections may also be made up by means of a

Fig. 4.6 W 14 steel column in substructure of a multistory building .(Courtesy Am. Inst. of Steel Construction.)

cover-plated W member, as shown in Fig. 4.5(h), or an S (standard beam) section together with two channel shapes. The latter section might be desirable for a very long member with not so great loads for which the cross section should be spread as much as possible to reduce the l/r ratio. Although Figs. 4.5(h) and (i) are shown as welded and bolted (or riveted), respectively, either fabrication method could be used on either section.

Figure 4.6 shows a heavy W 14 column section shortly after erection in one of the lower floors of an office building. The heavy bolted column splice has been completed at the lower end, and the splice plates at the upper end are ready to receive the next two-story column segment. Note the beam-connecting plates that are welded to the columns and field bolted to the beams.

(g) *Columns with Lacing, Battens, or Perforated Cover Plates*

In a situation where a very long column in required, it may be necessary to spread the cross section to a degree that makes a laced column economical, as shown for example in Fig. 4.7(a). Before the advent of the rolled W sections such laced members were also used for more usual lengths found in bridges and buildings, where they may still be seen in older structures. Laced columns are used today in derrick booms and TV or radio towers. In such applications the four-angle section shown in Fig. 4.7(a) together with the lacing bars may be replaced by solid round bars with welded end connections, which help reduce wind and icing loads on exposed members.

Lacing bars carry no column load, but they do perform the following functions:

1. They hold the component parts of the laced column in position so as to maintain the shape of the overall column cross section. For the same purpose, cross bracing in a plane normal to the column axis must be provided intermittently, as shown in the sectional view at the top of Fig. 4.7(a).

2. The lacing provides lateral support for the component column segments at each connecting point. For example, in Fig. 4.7(a), the l_0/r_0 of each individual angle, between support points, must be less than the overall l/r of the whole member. The r_0 of the individual angle should be the minimum value, as tabulated in the AISCM for the inclined z-z axis.

3. The lacing acts as a web replacement to resist shear and provide for the corresponding transfer of longitudinal variations of stress in the component longitudinal elements. In a centrally loaded column, shear force arises from

Fig. 4.7   Types of built-up columns.

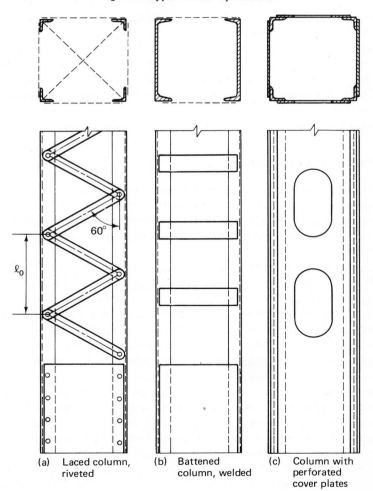

(a) Laced column, riveted

(b) Battened column, welded

(c) Column with perforated cover plates

accidental end eccentricity of the load and from curvature of the member under load. The shear force provides the basis for lacing bar design, and under the AISCS, Sec. 1.18.2.6, the resultant shear for design is taken as 2 per cent of the axial load in the member. Lacing bars may be designed as secondary members, but must act either in tension or compression and be designed for both loading conditions. Complete design rules for built-up members, including laced members and their end tie plates, are provided in Sec. 1.18, AISCS.

Battened columns [Fig. 4.7(b)] are not covered by the AISCS, but are used occasionally. They present the same design problems as do laced columns, but in comparison with the truss action of the lacing members, the battens resist shear by the less effective and more complex continuous-frame action.

Tie plates at the ends of both laced and battened members are particularly important to distribute the applied end loads. In weakly battened columns they may add materially to the overall column strength. For a more complete coverage of design of both laced and battened columns, reference should be made to Chapter 12 of the CRC *Guide* (third edition, in preparation).

Columns with perforated cover plates [Fig. 4.7(c)] are used chiefly in bridge construction. See also Fig. 2.6. The net section of such plates may be included in the column area, and they resist shear more effectively than either the laced or battened column. The perforations are provided primarily for drainage in exposed location and to provide access for cleaning and painting the interior surfaces. Simple design rules for such members are provided in all modern bridge specifications.

4.5 WIDTH–THICKNESS RATIOS

Width–thickness limitations are established to ensure that overall column buckling rather than local buckling governs the allowable design stress. When the limitations are not exceeded, the full cross section of the column may be considered to be effective. Limits on width–thickness ratios have been discussed in Sec. 4.4 for the case of the angle and tee section.

Limitations are as specified in Sec. 1.9, pages 5-25 and 5-26, of the AISCS, and the same topic should be studied in the AISCS pages 5-140 to 5-142. Width–thickness limits are established under two broad categories, *unstiffened elements* and *stiffened elements*, as defined in AISCS Sec. 1.9, and illustrated herein in Fig. 4.8. For equal width–thickness ratios, a stiffened element is much more effective than an unstiffened element, and much greater ratio limits are allowed for the stiffened element. As the yield stress increases, a more stocky element (smaller width–thickness ratio) is required to prevent premature local buckling under the increased allowable stress.

A complete tabulation of numerical values of width–thickness limits for the various yield stresses will be found in the AISCS, Appendix A, pages 5-72 to 5-75.

When a thin-walled member does double duty both as a column and partition, it may be desirable to exceed the width–thickness limits. Such members may be used provided a "reduced effective width" and/or reduced allowable stresses are employed, as covered in Appendix C of the AISCS, pages 5-115 to 5-117, or in the AISI specifications. Reference also may be made to Chapter 9 of the CRC *Guide* (third edition).

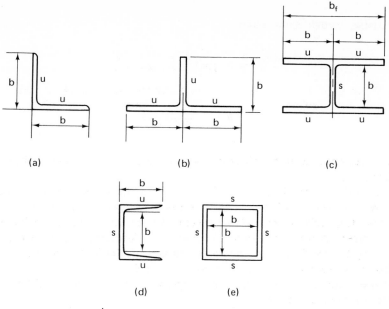

$\dfrac{b}{t}$ = width/thickness ratio of element

u = unstiffened element

s = stiffened element

Fig. 4.8 Stiffened and unstiffened plate elements of structural shapes as defined for AISCS Sec. 1.9 width–thickness limitations.

4.6 COLUMN BASE PLATES AND SPLICES

Columns on footings must be provided with base plates to distribute load to the masonry within the allowable bearing capacity of the concrete. For very heavy columns in tall buildings, individual base plates may not be sufficient, and a grillage system may be required. For information on base-plate design, including design examples and tables of available sizes, reference should be made to the AISCM, pages 3-95 to 3-103. Suggested general details for base plates are also provided on pages 4-105 and 4-106. Riveted, bolted, and welded column splice details are illustrated on pages 4-107 to 4-110 of the AISCM (see also Fig. 4.6).

4.7 ALLOWABLE COMPRESSION STRESS

Insofar as general buckling under axial load governs the design selection of a steel column, the allowable stress at working load is defined by the AISCS, Sec. 1.5.1.3, page 5-16, and is based on the column-strength curves, Eqs. (4.4) and (4.3), divided by an appropriate factor of safety (*FS*).

(a) When $Kl/r < C_c$, apply the CRC basic column-strength curve, Eq. (4.4); then the allowable compression stress is

$$F_a = \left[1 - \frac{(Kl/r)^2}{2C_c^2}\right]\frac{F_y}{FS} \qquad (4.7)$$

where
$$FS = \frac{5}{3} + \frac{3}{8C_c}\left(\frac{Kl}{r}\right) - \frac{1}{8C_c^3}\left(\frac{Kl}{r}\right)^3 \tag{4.8}$$

(b) When $C_c < Kl/r < 200$, apply the Euler formula, Eq. (4.3); then

$$F_a = \frac{\pi^2 E}{FS(Kl/r)^2} \tag{4.9}$$

where
$$FS = \tfrac{23}{12} = 1.92 \tag{4.10}$$

It is noted that the allowable compression stress is independent of the yield point when $Kl/r > C_c$.

At Kl/r of zero, the factor of safety of 1.67 is the same in compression as for tension. With increasing Kl/r, the factor of safety increases toward 1.92. This allows for uncertainties such as unavoidable eccentricity, residual stress, crookedness, and, for very slender columns, the increased sensitivity to uncertainties in the evaluation of effective length. The foregoing allowable stresses are limited to members that meet the width–thickness requirements of AISCS, Sec. 1.9.

For the bracing and secondary members, when l/r is between 120 and 200, the allowable compression stress F_{as} is

$$F_{as} = \frac{F_a}{1.6 - (l/200r)} \tag{4.11}$$

Allowable compression stresses for steels having yield stresses of 36, 42, 45, 50, 55, 60, 65, 90, and 100 ksi are tabulated in Table 1-36 in Appendix A of the AISCS, pages 5-84 to 5-92.

The following flow chart systematizes the use of the AISCS for the selection of allowable column stress under axial load. It will also be used as an adjunct to Chapter 5 for the design of beam-columns, as the value of F_a enters into the AISCS interaction formula for beam-column design.

In addition to the AISCS and Commentary, the AISCM notes on column design and accompanying examples, pages 3-3 to 3-7, should be studied.

The manual solution of a column design problem, using either slide rule or desk computer, illustrates very well the basic difference between a problem of *analysis* of a given structure and the problem of *design*, for which the rules (specifications) are given and the structure is the end product. In column design the allowable stress is a function of the slenderness ratio and is therefore generally unknown in advance of a trial design selection. Exceptions to this statement are represented by the tube, of a given diameter or size, and the built-up column of four angles. In these cases the suitable general dimensions of the cross section can usually be established in advance, thus pinpointing the slenderness ratio, and the final design is arrived at by simply changing the wall or angle thickness to provide the required area that is equal to the load divided by the allowable stress. Similarly, in heavy tier building construction, using W 14 sections, the radius of gyration of the cross section will be known approximately in advance. But in many cases it may be necessary to simply "guess" at some preliminary trial selection and arrive at the final design by means of one or two additional trial selections. One may either guess the radius of gyration, or, more directly, guess the allowable stress. The analysis of such a trial selection will quickly lead to a much better trial. Most of the examples presented herein involve the check of a final trial selection—in each case the reader should ask himself: How best could this trial have been arrived at?

4.8 FLOW CHARTS

Flow Chart 4.1

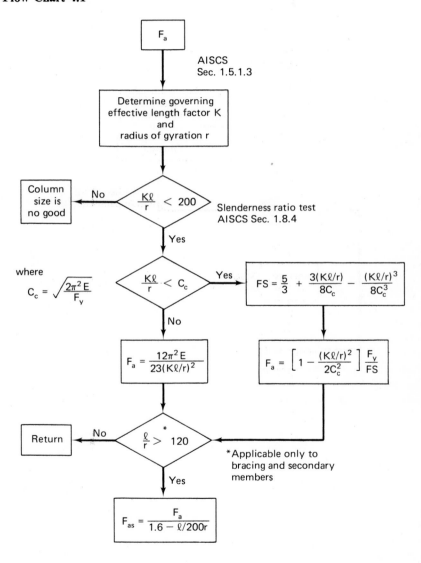

Flow Chart 4.2

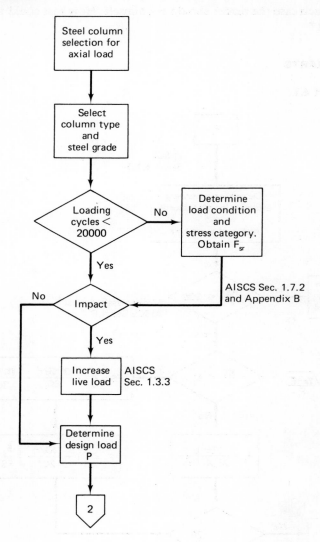

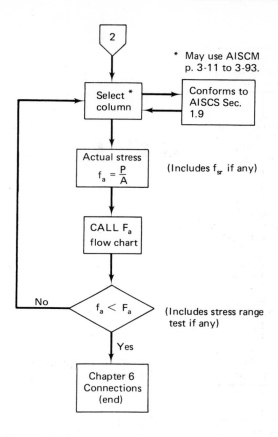

* May use AISCM
  p. 3-11 to 3-93.

(Includes f_{sr} if any)

(Includes stress range
 test if any)

4.9 ILLUSTRATIVE EXAMPLES

Example 4.1

A portion of a trussed TV antenna tower as shown has main longitudinal elements that carry an axial load P (270 kips) and are laterally braced at 6-ft intervals. No advantage will be taken of rotation restraint provided by lateral bracing at ends. The dashed line shown indicates a natural buckling mode of the main compression elements at failure. Select a solid round bar for the main compression elements to satisfy the AISCS. Use A36 steel.

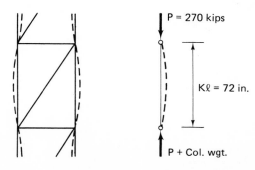

Solution

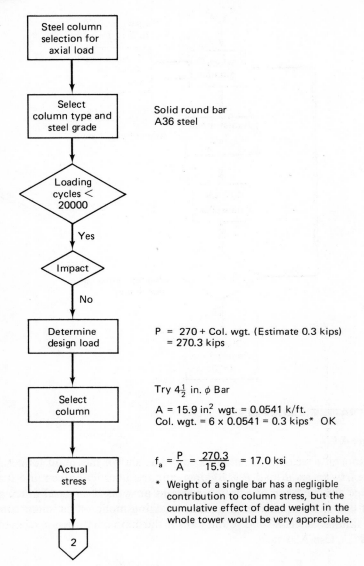

Steel column selection for axial load

Select column type and steel grade — Solid round bar A36 steel

Loading cycles < 20000

Yes

Impact

No

Determine design load

$P = 270 + $ Col. wgt. (Estimate 0.3 kips)
$= 270.3$ kips

Select column

Try $4\frac{1}{2}$ in. ϕ Bar

$A = 15.9$ in.^{2} wgt. $= 0.0541$ k/ft.
Col. wgt. $= 6 \times 0.0541 = 0.3$ kips* OK

Actual stress

$f_a = \dfrac{P}{A} = \dfrac{270.3}{15.9} = 17.0$ ksi

* Weight of a single bar has a negligible contribution to column stress, but the cumulative effect of dead weight in the whole tower would be very appreciable.

2

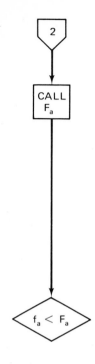

See F_a flow chart

Effective length factor K = 1.0 (See Table 4.1)

Radius of gyration $r = \dfrac{d}{4} = \dfrac{4.5}{4} = 1.125$ in.

Slenderness ratio $\dfrac{K\ell}{r} = \dfrac{1.0(6 \times 12)}{1.125} = 64.0$

$C_c = \sqrt{2\pi^2 E/F_y} = \sqrt{2\pi^2 \times 29000/36} = 126.1$

(see AISCS Table 1-B, p. 5-93)

$\dfrac{K\ell}{r} < C_c$, then

$FS = \dfrac{5}{3} + \dfrac{3 \times 64}{8 \times 126.1} - \dfrac{64^3}{8 \times 126.1^3}$

$\quad = 1.844$

$F_a = [1 - \dfrac{64^2}{2 \times 126.1^2}] \dfrac{36}{1.844} = 17.04$ ksi

(See AISCS Table 1-36, p. 5-84)

Stress test OK

USE a Bar $4\frac{1}{2}\,\phi$

Note: In computer aided design, formulas for C_c, F_a,
etc. would be part of the program. In manual
design, maximum use would be made of charts and
tables, as referred to above. Calculation
details as given above are to provide a complete
illustration of specification requirements and
are not typical of actual design computations.

Example 4.2

For an unbraced and effectively hinged-end length of 8 ft, design a steel pipe column for an axial load of 50 kips, using A36 steel, AISCS, and limiting the selection to standard pipe sizes.

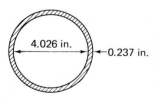

4.026 in.   0.237 in.

Solution

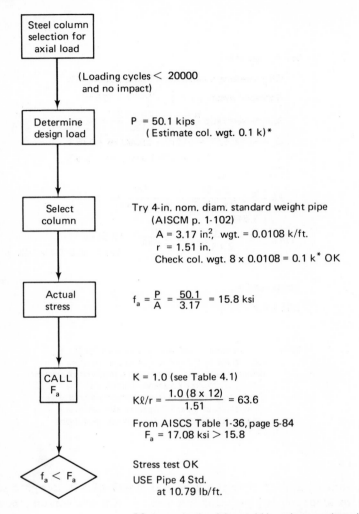

Steel column selection for axial load

(Loading cycles $<$ 20000 and no impact)

Determine design load

P = 50.1 kips
(Estimate col. wgt. 0.1 k)*

Select column

Try 4-in. nom. diam. standard weight pipe
(AISCM p. 1-102)
A = 3.17 in^2, wgt. = 0.0108 k/ft.
r = 1.51 in.
Check col. wgt. 8 x 0.0108 = 0.1 k* OK

Actual stress

$f_a = \dfrac{P}{A} = \dfrac{50.1}{3.17} = 15.8$ ksi

CALL F_a

K = 1.0 (see Table 4.1)

$K\ell/r = \dfrac{1.0\,(8 \times 12)}{1.51} = 63.6$

From AISCS Table 1-36, page 5-84
F_a = 17.08 ksi $>$ 15.8

$f_a < F_a$

Stress test OK
USE Pipe 4 Std.
at 10.79 lb/ft.

*Column dead weight could have been neglected.

Example 4.3

Design a square structural tube for the same conditions as in Ex. 4.2.

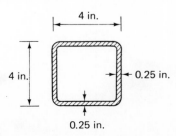

Solution

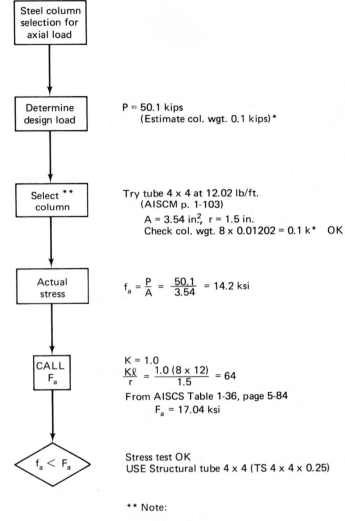

P = 50.1 kips
    (Estimate col. wgt. 0.1 kips)*

Try tube 4 x 4 at 12.02 lb/ft.
    (AISCM p. 1-103)
        A = 3.54 in.^{2}, r = 1.5 in.
        Check col. wgt. 8 x 0.01202 = 0.1 k*   OK

$f_a = \dfrac{P}{A} = \dfrac{50.1}{3.54} = 14.2$ ksi

K = 1.0
$\dfrac{K\ell}{r} = \dfrac{1.0\,(8 \times 12)}{1.5} = 64$
From AISCS Table 1-36, page 5-84
        $F_a = 17.04$ ksi

Stress test OK
USE Structural tube 4 x 4 (TS 4 x 4 x 0.25)

** Note:

Check width/thickness ratio
(AISCS Sec. 1.9.2.2, p. 5-74) :
        $\dfrac{b}{t} = \dfrac{4}{0.25} = 16 < 39.7$ OK

* Dead weight could have been neglected.

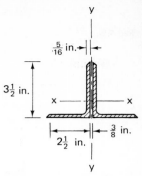

Example 4.4

Design a double-angle strut for the same conditions as in Ex. 4.2 with the added stipulation that special attention must be given to end connections to bring the axial load uniformly into both legs at each end.

Solution

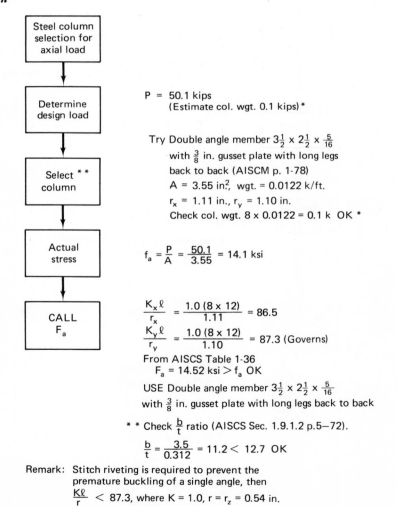

$P = 50.1$ kips
    (Estimate col. wgt. 0.1 kips)*

Try Double angle member $3\frac{1}{2} \times 2\frac{1}{2} \times \frac{5}{16}$
with $\frac{3}{8}$ in. gusset plate with long legs
back to back (AISCM p. 1-78)
$A = 3.55$ in.², wgt. = 0.0122 k/ft.
$r_x = 1.11$ in., $r_y = 1.10$ in.
Check col. wgt. $8 \times 0.0122 = 0.1$ k OK *

$$f_a = \frac{P}{A} = \frac{50.1}{3.55} = 14.1 \text{ ksi}$$

$$\frac{K_x \ell}{r_x} = \frac{1.0\,(8 \times 12)}{1.11} = 86.5$$

$$\frac{K_y \ell}{r_y} = \frac{1.0\,(8 \times 12)}{1.10} = 87.3 \text{ (Governs)}$$

From AISCS Table 1-36
    $F_a = 14.52$ ksi $> f_a$ OK

USE Double angle member $3\frac{1}{2} \times 2\frac{1}{2} \times \frac{5}{16}$
with $\frac{3}{8}$ in. gusset plate with long legs back to back

** Check $\frac{b}{t}$ ratio (AISCS Sec. 1.9.1.2 p.5−72).

$$\frac{b}{t} = \frac{3.5}{0.312} = 11.2 < 12.7 \text{ OK}$$

Remark: Stitch riveting is required to prevent the premature buckling of a single angle, then
$\frac{K\ell}{r} < 87.3$, where $K = 1.0$, $r = r_z = 0.54$ in.

$$\ell < \frac{87.3r}{K} = \frac{87.3 \times 0.54}{1.0} = 47.2 \text{ in.}$$

USE spacing of 3 ft 6 in. c. to c. stitch rivets.
    *Dead weight could have been neglected.

Example 4.5

Design a structural tee column for the same conditions as in Ex. 4.2.

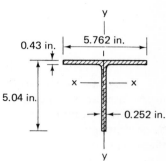

Solution

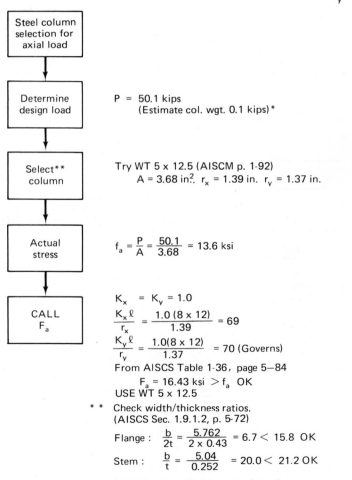

Steel column selection for axial load

Determine design load

P = 50.1 kips
    (Estimate col. wgt. 0.1 kips)*

Select column**

Try WT 5 x 12.5 (AISCM p. 1-92)
    A = 3.68 in.^{2}, r_x = 1.39 in. r_y = 1.37 in.

Actual stress

$$f_a = \frac{P}{A} = \frac{50.1}{3.68} = 13.6 \text{ ksi}$$

CALL F_a

$K_x = K_y = 1.0$

$$\frac{K_x \ell}{r_x} = \frac{1.0 (8 \times 12)}{1.39} = 69$$

$$\frac{K_y \ell}{r_y} = \frac{1.0(8 \times 12)}{1.37} = 70 \text{ (Governs)}$$

From AISCS Table 1-36, page 5—84
    F_a = 16.43 ksi > f_a OK
USE WT 5 x 12.5

**   Check width/thickness ratios.
    (AISCS Sec. 1.9.1.2, p. 5-72)

Flange : $\dfrac{b}{2t} = \dfrac{5.762}{2 \times 0.43} = 6.7 < 15.8$ OK

Stem : $\dfrac{b}{t} = \dfrac{5.04}{0.252} = 20.0 < 21.2$ OK

*Dead weight could have been neglected.

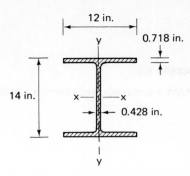

Example 4.6

Design a wide-flange (W) column 20 ft in length to support an axial load of 410 kips in the interior of a building. The column base is rigidly fixed to the footing and the top of the column is rigidly framed to very stiff girders. Assume bracing is provided to prevent sidesway in the weak deflection plane of the column, but the sidesway in the strong plane is not prevented. Select an economical W shape to satisfy the AISCS. Use A36 steel.

Solution

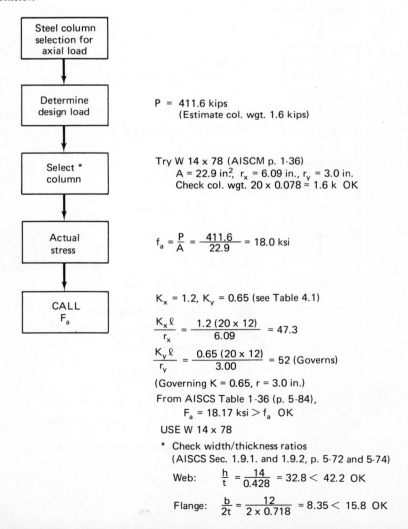

P = 411.6 kips
   (Estimate col. wgt. 1.6 kips)

Try W 14 x 78 (AISCM p. 1-36)
   A = 22.9 in², r_x = 6.09 in., r_y = 3.0 in.
   Check col. wgt. 20 x 0.078 = 1.6 k OK

$$f_a = \frac{P}{A} = \frac{411.6}{22.9} = 18.0 \text{ ksi}$$

K_x = 1.2, K_y = 0.65 (see Table 4.1)

$$\frac{K_x \ell}{r_x} = \frac{1.2\,(20 \times 12)}{6.09} = 47.3$$

$$\frac{K_y \ell}{r_y} = \frac{0.65\,(20 \times 12)}{3.00} = 52 \text{ (Governs)}$$

(Governing K = 0.65, r = 3.0 in.)
From AISCS Table 1-36 (p. 5-84),
   F_a = 18.17 ksi > f_a OK
USE W 14 x 78

* Check width/thickness ratios
  (AISCS Sec. 1.9.1. and 1.9.2, p. 5-72 and 5-74)

Web: $\dfrac{h}{t} = \dfrac{14}{0.428} = 32.8 < 42.2$ OK

Flange: $\dfrac{b}{2t} = \dfrac{12}{2 \times 0.718} = 8.35 < 15.8$ OK

Example 4.7†

Same as Ex. 4.6, except the top of the column is rigidly framed to two-way floor beams, as shown in the figure. Determine the axial load capacity of the selected W 14 × 78 column in the previous example. Assume all stories are spaced at 20 ft and adjacent stories have the same column section. Use AISCS and A36 steel.

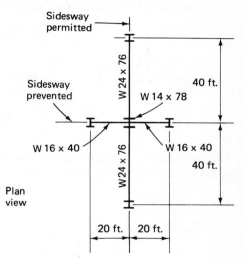

Member properties:

W 14 × 78 column:	$A = 22.9$ in.^{2}, $I_x = 851$ in.^{4}, $I_y = 207$ in.^{4}
	$r_x = 6.09$ in. $r_y = 3.00$ in.
W 24 × 76 beam:	$I_x = 2100$ in.^{4}
W 16 × 40 beam:	$I_x = 517$ in.^{4}

Solution

Determine the governing slenderness ratio:

Case 1

Major x axis (sidesway permitted): At column top,

$$G_A = \frac{\sum I_c/L_c}{\sum I_g/L_g}$$
$$= \frac{(851/20) + (851/20)}{(2100/40) + (2100/40)} = 0.813$$

At column bottom $G_B = 1.0$ (AISCS Commentary, page 5-139) for column base rigidly attached to a footing.

Apply Fig. 4.4(b) for $G_A = 0.813$, $G_B = 1.0$; then

†See AISCM, pages 3-4 and 3-5, for explicit illustration of the use of the alignment chart prior to study of this example.

$$K_x = 1.29$$
$$\frac{K_x l}{r_x} = \frac{1.29(20 \times 12)}{6.09} = 50.8$$

Case 2

Minor y axis (sidesway prevented):† At column top,.

$$G_A = \frac{(207/20) + (207/20)}{(517/20) + (517/20)} = 0.402$$

At column bottom $G_B = 1.0$.
Apply Fig. 4.4(a) for $G_A = 0.402$, $G_B = 1.0$; then

$$K_y = 0.717$$
$$\frac{K_y l}{r_y} = \frac{0.717(20 \times 12)}{3.00} = 57.3$$

Therefore, the slenderness ratio is governed by case 2 for which the buckling will occur in the column about the minor y axis:

$$\frac{Kl}{r} = 57.3$$

Therefore

$$F_a = 17.68 \text{ ksi} \qquad \text{(from AISCS, Table 1-36)}$$

The axial load capacity of the W 14 × 78 column is

$$P = F_a A = 17.68 \times 22.9 = 405 \text{ kips}$$

PROBLEMS

4.1. Determine the diameter of a solid round A36 steel bar, 9 ft in length, that will support an axial force of 160 kips. Choose an available size. Lateral support is assumed at each end. The ends are assumed to be hinged; hence the effective column length is the same as the actual length.

4.2. Select standard A36 steel shapes having an area as near as possible to that of the requirement determined for Problem 4.1 and compare the column load capacities for the following:
  (a) An W 8 section.
  (b) A single angle with equal legs.

†Since the AISCS Commentary does not include the alignment chart for "sidesway prevented," one might conservatively take $K_y = 0.8$, as recommended in Table 4.1 (AISCS Commentary, Table Cl.8.1, page 5-138), for the case with fixed base and hinged top. In that case,

$$\frac{K_y l}{r_y} = \frac{0.80 \times 240}{3.00} = 64.0$$

and
$$F_a = 17.04 \text{ ksi}$$
$$P = 17.04 \times 22.9 = 390 \text{ kips}$$

(c) A double angle with long legs spaced $\frac{3}{8}$ in. apart.

(d) A structural tee.

(e) A round pipe.†

(f) A square structural tube.†

4.3. Assume for the purposes of this problem that the total weight of a 30-story building and its contents (vertical live load) averages 180 lb/ft² of floor area per floor. For a floor plan 90 ft by 180 ft, encompassing 40 columns, make a preliminary first-floor trial-column-size selection for an effective column length of 14 ft. Columns are assumed to be axially loaded and the choice can be made with the aid of the AISCM column load tables—but the actual capacity of the column for axial load should be checked by the AISCS formula. Assume equal distribution of load to all columns.

4.4. Rework Problem 4.3 on the assumption that the columns will be braced laterally in their weak direction.

4.5. A WT 6 × 20 is welded to a PL $\frac{3}{4}$ × 12 as shown. The overall length is 20 ft and the column is fixed at the base, hinged at the top. Use recommended effective length factor from Table 4.1. Determine the allowable column load for (a) A36 steel, and (b) A steel with $F_y = 50$ ksi.

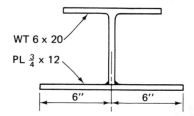

WT 6 x 20

PL $\frac{3}{4}$ x 12

6″   6″

4.6. A hinged end column 36 ft in length is supported in its weak direction only, at the third points, in which case it may be assumed to behave as if the effective length for buckling about the y-y axis is 12 feet. Select a section to carry a load of 90 kips, using A36 steel, and providing as near equal slenderness ratios as possible about the two axes.

4.7. Considering the column cross section, length, local details, and bracing, as shown, what is the permissible load "P" utilizing A36 steel? See AISCS Sec. 1.18.2.6 for design assumptions and select an adequate size for the lacing bars, which may be assumed to have adequate welded end connections and which need to be selected on the basis of adequate compressive strength. See figure on next page.

4.8. Design a column 60 ft in length to carry a load of 600 kips without any intermediate lateral support. Use a cross section consisting of four corner angles with double lacing similar to that used in Problem 4.7 on all four sides. (See sketch of cross section on next page.) Design all details, including size of lacing, except for welded connections between lacing and angles. Follow requirements of AISCS Sec. 1.18.2.6 and use a steel with $F_y = 50$ ksi.

---

†Use the largest available outside dimension for the given area.

Note: Table 1-36, page 5-84 of AISCM may be used.

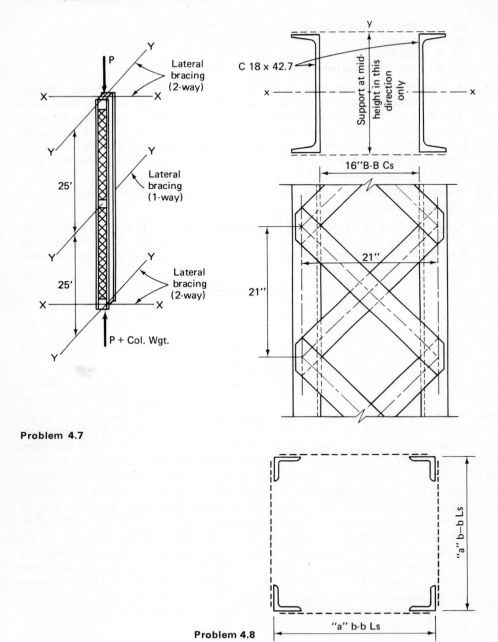

Problem 4.7

Problem 4.8

4.9. Rework Example 4.6, changing steel from A36 to $F_y = 50$ ksi.

4.10. Rework Example 4.7, changing steel from A36 to $F_y = 50$ ksi.

5

COLUMNS
UNDER COMBINED STRESS[†]

5.1 INTRODUCTION

In Chapter 3 we treated steel beam design and in Chapter 4 the design of axially loaded columns. In actual structures, most columns, in addition to axial load, must support lateral loads and/or transmit moments between their ends, and are thus subjected to combined stress due both to axial load and moment. Such members are sometimes termed beam-columns. The end moments may be caused by continuous-frame action and/or by the effective eccentricity of the longitudinal loads. For example, columns in a tall building frame, in addition to the live and dead load of the structure above any given level, must transmit bending moments that result from wind load or lateral inertia forces due to earthquake. In addition, they must resist the end moments introduced through the continuous frame action of the loaded adjacent connecting beams. In a tall building the moments induced by the sway deflection of the structure must also be included.

When a column is part of a frame, the ideal solution would be to determine the strength of the complete structure and use this as a basis for design. There is a trend toward such a design procedure, but at this time the traditional method of isolating the individual member as a basis for design prevails. The design may then proceed along one of the following three paths:

[†]The application of AISCS, Sec. 1.6.2, to the tension member under combined stress is also illustrated in Example 5.5.

111

1. The load at which the maximum stress reaches the yield stress is determined and that load is divided by a factor of safety to give an allowable load. Moment due to deflection is included.

2. The AISCS interaction formula may be used. This provides for an empirical transition in the member selection, in accordance with AISCS requirements, from the beam (as the column load approaches zero) to the axially loaded column (as the bending moment approaches zero).

3. The AISCS plastic-design procedure, also using an interaction formula similar to that mentioned in item 2, will be used if this alternative design method is adopted for continuous steel frame design. The principal difference is that under item 2 the moments are determined by elastic frame analysis (such as the moment-distribution method), whereas in plastic design the bending moments are distributed as they would be when the frame is at the verge of collapse.

5.2 ALLOWABLE-STRESS DESIGN

A very short and stocky beam-column, one, say, in which Kl/r is less than 15,† may be designed very simply and quite conservatively on the basis of full maximum allowable stress. In such a short member the additional bending moment due to deflection may be neglected. The designation "stocky" is important, because even a short member is subject to failure by local buckling and must meet the width–thickness requirements for compression members. The maximum stress is equal to the sum of the average stress due to axial load plus the compressive stress added by bending moment:

$$f_{\max} = \frac{P}{A} + \frac{M_x c_x}{I_x} \tag{5.1}$$

A more direct selection of a short beam-column is made possible by rewriting Eq. (5.1) along the following lines. I_x may be replaced by Ar_x^2 and, as shown in Fig. 5.1, the moment M may be replaced by its static equivalent, Pe_x; thus

$$f_{\max} = \frac{P}{A}\left(1 + \frac{e_x c_x}{r_x^2}\right) \tag{5.2a}$$

$$\frac{P}{A} = \frac{f_{\max}}{1 + (e_x c_x / r_x^2)} \tag{5.2b}$$

Thus, by means of Eq. (5.2b), the average allowable column stress may be obtained very simply by substituting the full allowable stress for a zero length column ($0.6F_y$) in place of $f_{\max}$. For the stipulated limit on l/r, the actual column stress will be no more than about 3 per cent less than this value. If the section is "compact," the bending part of the stress would have a higher allowable value of $0.66F_y$,—a feature that is introduced by the interaction formula to be discussed in the next section.

If there are components of moment about two axes of the cross section, Eq. (5.2b) may be modified to the following:

†For $F_y > 36$ ksi, use $90/\sqrt{F_y}$ in place of 15.

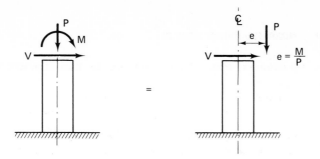

Fig. 5.1 Equivalent loads applied to a short beam-column.

$$\frac{P}{A} = \frac{f_{max}}{1 + (e_x c_x / r_x^2) + (e_y c_y / r_y^2)} \tag{5.3}$$

In applying either Eq. (5.2b) or Eq. (5.3) to design, values of c and r about each axis may be approximated by reference to the design information in the AISCM for a particular series of sections. A trial selection is then made on the basis of the required P/A. If satisfactory, the trial values of c and r may then be replaced by the actual values for the selected section and the design rechecked.

Although Eq. (5.3) may be further modified† to adapt it to longer columns by including the effect of deflection, attention is now turned to the AISCS interaction formula procedure, which is currently widely used in design of beam-columns.

5.3 DESIGN BY USE OF INTERACTION FORMULAS

Using AISCS notation, Eq. (5.1) may be written

$$f_{max} = f_a + f_{bx}$$

or, if there are moments about both axes of the cross section,

$$f_{max} = f_a + f_{bx} + f_{by}$$

Dividing both sides of the foregoing by f_{max},

$$1 = \frac{f_a}{f_{max}} + \frac{f_{bx}}{f_{max}} + \frac{f_{by}}{f_{max}} \tag{5.4}$$

Equation (5.4) is now in the form termed an interaction formula, but for design application it can be improved by introducing the allowable stress applicable to each of the three terms on the right-hand side of Eq. (5.4) in place of f_{max}. Then if any two of the terms on the right side become zero, the correct specified allowable stress is approached in the limit as the beam-column becomes an axially loaded column or a beam bent about either of the two axes. With such a modification Eq. (5.4) becomes the AISCS interaction equation for design under combined stress for the case where f_a/F_a is less than 0.15;

$$\frac{f_a}{F_a} + \frac{f_{bx}}{F_{bx}} + \frac{f_{by}}{F_{by}} \leq 1.0 \qquad \text{[AISCS (1.6-2)]}$$

When f_a/F_a is greater than 0.15, additional bending moments due to deflection ($P\delta_x$ and $P\delta_y$) may contribute significantly to combined stress. (See Fig. 2.2.) The

†See *The USS Steel Design Manual* for this procedure.

additional bending moment may be approximated by multiplying both f_{bx} and f_{by} by an *amplification factor*,

$$\frac{C_m}{1 - (f_a/F'_e)}$$

where F'_e is the Euler buckling stress [Eq. (4.3)] divided by $\frac{23}{12}$, or 1.92, which is the AISCS factor of safety for the very long column with KL/r greater than C_c. C_m is a correction factor to the standard amplification, usually about 1, which adjusts for variations in the distribution of bending moment along the member, as covered by Sec. 1.6.1, page 5-23, of the AISCS. At this point, pages 5-130 to 5-134 of the AISCS Commentary should be studied relative to AISCS, Sec. 1.6.1.

With the introduction of the amplification factors, AISCS Eq. (1.6-2) is modified so as to be applicable to the case when f_a/F_a is greater than 0.15, as follows:

$$\frac{f_a}{F_a} + \frac{C_{mx}f_{bx}}{\left(1 - \dfrac{f_a}{F'_{ex}}\right)F_{bx}} + \frac{C_{my}f_{by}}{\left(1 - \dfrac{f_a}{F'_{ey}}\right)F_{by}} \leq 1.0 \qquad \text{[AISCS (1.6-la)]}$$

and, in addition, at points braced in the plane of bending,

$$\frac{f_a}{0.6F_y} + \frac{f_{bx}}{F_{bx}} + \frac{f_{by}}{F_{by}} \leq 1.0 \qquad \text{[AISCS (1.6-lb)]}$$

where the subscripts x and y indicate the axis of bending about which a particular stress or design property applies, and

F_a = allowable axial compression stress for axial force alone
F_b = allowable bending stress for bending moment alone
$$F'_e = \frac{12\pi^2 E}{23(Kl_b/r_b)^2} \qquad \text{(see AISCS, Table 2, page 5-94)}$$

where l_b is the actual unbraced length in the plane of bending and r_b is the corresponding radius of gyration. K is the effective length factor in the plane of bending. As in the case of F_a, F_b, and 0.6 F_y, F'_e may be increased 33.3 per cent for wind and seismic loads in accordance with Sec. 1.5.6.

f_a = actual axial compression stress
f_b = actual maximum fiber compressive bending stress

C_m is a coefficient as follows:

1. For compression members in frames subject to sidesway, $C_m = 0.85$.

2. For restrained compression members in frames braced against sidesway and not subject to transverse loading between their supports in the plane of bending,

$$C_m = 0.6 - 0.4\frac{M_1}{M_2}$$

but not less than 0.4, where M_1/M_2 is the ratio of the smaller to larger moments at the ends of that portion of the member unbraced in the plane of bending under consideration. M_1/M_2 is positive when the member is bent in reverse curvature and negative when it is bent in single curvature.

3. For compression members in frames braced against sidesway in the plane of loading and subjected to transverse loading between their supports, the value of C_m may be determined by rational analysis. However, in lieu of such analysis the following values may be used: (a) for members whose ends are restrained, $C_m = 0.85$; (b) for members whose ends are unrestrained, $C_m = 1.0$.

5.4 EQUIVALENT AXIAL COMPRESSION LOAD

Both the preliminary selection and final design check of a beam-column, proportioned by the AISCS interaction formulas, may be expedited by conversion to an equivalent axial load, thereby making use of the tabulated loads provided on pages 3-12 to 3-56 of the AISCM. Any one of the AISCS interaction formulas (1.6-1a), (1.6-1b), or (1.6-2) can be generalized as follows:†

$$\alpha_a \frac{f_a}{F_a} + \alpha_{bx} \frac{f_{bx}}{F_{bx}} + \alpha_{by} \frac{f_{by}}{F_{by}} \le 1.0$$

Multiplying both sides of the inequality by $AF_{a'}$

$$\alpha_a A f_a + \alpha_{bx} \frac{f_{bx}}{F_{bx}} AF_a + \alpha_{by} \frac{f_{by}}{F_{by}} AF_a \le AF_a$$

AF_a would be the allowable load if the column were axially loaded. Thus the sum of the three terms on the left of the foregoing inequality can be thought of as an equivalent axial load and used as an entry into the AISCM column-selection tables. The use of the AISCM is further facilitated by rewriting the previous equation in terms of M_x and M_y, and by introducing coefficients B_x and B_y, which are also tabulated in the AISCM:

$$Af_{bx} = \frac{M_x A}{S_x} = M_x B_x, \qquad Af_{by} = \frac{M_y A}{S_y} = M_y B_y$$

Noting that Af_a is the actual column load, the equivalent load equation, in general terms, becomes

$$P_{eq} = \alpha_a P + \alpha_{bx} \frac{F_a}{F_{bx}} B_x M_x + \alpha_{by} \frac{F_a}{F_{by}} B_y M_y \qquad (5.5)$$

We may now rewrite AISCS Eqs. (1.6-1a), (1.6-1b), and (1.6-2) as follows: Modified AISCS Eq. (1.6-1a), for $f_a/F_a > 0.15$:

$$P_{eq} = P + B_x M_x C_{mx} \frac{F_a}{F_{bx}} \frac{1}{1 - (P/AF'_{ex})} + B_y M_y C_{my} \frac{F_a}{F_{by}} \frac{1}{1 - (P/AF'_{ey})}$$

Although the foregoing may be used as it stands, the AISCM further simplifies

---

†α_a, α_{bx}, and α_{by} are simply coefficients that represent a collection of terms in any one of the three AISCS formulas. In the very short strut each $\alpha = 1$ [see Eq. (5.4)].

the amplification portions by the tabulation of additional section coefficients a_x and a_y. Introducing the formula for F'_e,

$$AF'_e = \frac{12\pi^2 EAr^2}{23(Kl)^2}$$

we let a (subscripts x and y temporarily omitted) be

$$a = \frac{12\pi^2 EAr^2}{23}$$

With the further simplification of AISCS Eq. (1.6-1a), all three modified formulas are now summarized:

P_{eq} = required tabular load

$$= P + \left[B_x M_x C_{mx} \left(\frac{F_a}{F_{bx}}\right) \left(\frac{a_x}{a_x - P(Kl)^2}\right) \right]$$

$$+ \left[B_y M_y C_{my} \left(\frac{F_a}{F_{by}}\right) \left(\frac{a_y}{a_y - P(Kl)^2}\right) \right] \qquad \text{Modified Formula (1.6-1a)}$$

P_{eq} = required tabular load

$$= P\left(\frac{F_a}{0.6F_y}\right) + \left[B_x M_x \left(\frac{F_a}{F_{bx}}\right) \right] + \left[B_y M_y \left(\frac{F_a}{F_{by}}\right) \right] \qquad \text{Modified Formula (1.6-1b)}$$

When $f_a/F_a \leq 0.15$

P_{eq} = required tabular load

$$= P + \left[B_x M_x \left(\frac{F_a}{F_{bx}}\right) \right] + \left[B_y M_y \left(\frac{F_a}{F_{by}}\right) \right] \qquad \text{Modified Formula (1.6-2)}$$

In Formula (1.6-1a), for the term $(Kl)^2$, K is the effective length factor and l is the actual unbraced length in the plane of bending.

Values for the components a_x and a_y, equal to $0.149 \times 10^6 Ar_x^2$ and $0.149 \times 10^6 Ar_y^2$, respectively, are listed at the bottom of the load tables.

The foregoing derivation and discussion of the modified interaction formulas should be supplemented by study of the foreword to the column load tables on pages 3-7 to 3-10 of the AISCM. One modification should be mentioned. The 1970 Supplement No. 1 to the AISCS authorized intermediate values for F_{by} between $0.60F_y$ and $0.75F_y$ when b/t_f lies between $95/\sqrt{F_y}$ and $52.2/\sqrt{F_y}$, respectively, as follows:

$$F_{by} = F_y \left[0.933 - 0.0035 \frac{b_f}{2t_f} \sqrt{F_y} \right] \qquad \text{[AISCS Eq. (1.5-5b)]†}$$

In using the modified formulas a quick trial selection may be made on the basis of the following crude approximation for the equivalent load:

$$P_{eq} = P + B_x M_x + B_y M_y \qquad (5.6)$$

Average values from the tables for B_x and B_y may be used for the initial trial selection. The use of Eq. (5.6) will result in an overestimate of the actual requirements; hence the initial trial should be taken so as to provide slightly less than the indicated value of P_{eq} by Eq. (5.6).

†Added in 1970.

5.5 FLOW CHARTS

Flow Chart 5.1

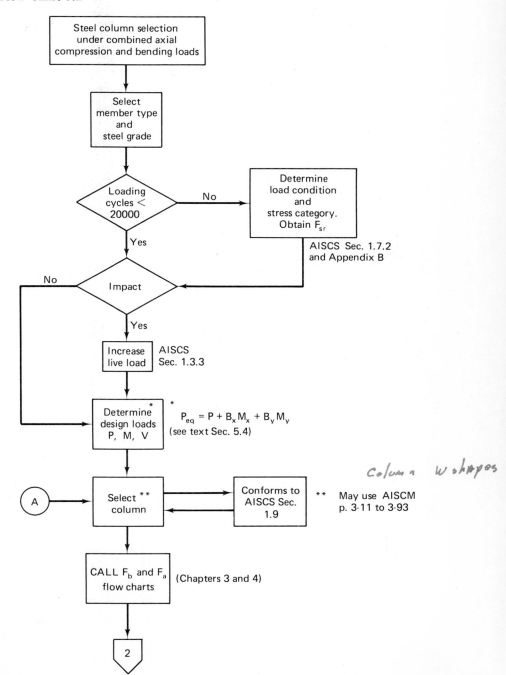

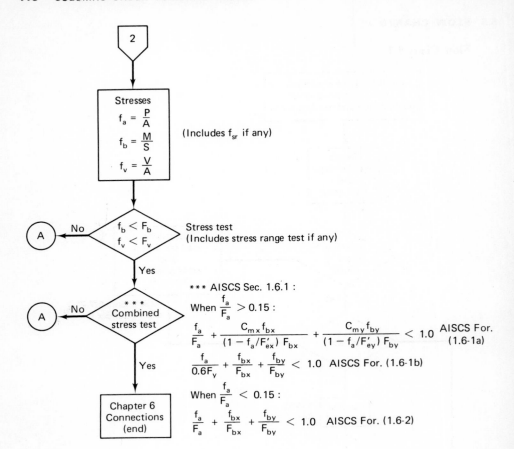

*** AISCS Sec. 1.6.1 :

When $\dfrac{f_a}{F_a} > 0.15$:

$$\frac{f_a}{F_a} + \frac{C_{mx} f_{bx}}{(1 - f_a/F'_{ex}) F_{bx}} + \frac{C_{my} f_{by}}{(1 - f_a/F'_{ey}) F_{by}} < 1.0 \quad \text{AISCS For. (1.6-1a)}$$

$$\frac{f_a}{0.6F_y} + \frac{f_{bx}}{F_{bx}} + \frac{f_{by}}{F_{by}} < 1.0 \quad \text{AISCS For. (1.6-1b)}$$

When $\dfrac{f_a}{F_a} < 0.15$:

$$\frac{f_a}{F_a} + \frac{f_{bx}}{F_{bx}} + \frac{f_{by}}{F_{by}} < 1.0 \quad \text{AISCS For. (1.6-2)}$$

Flow Chart 5.2

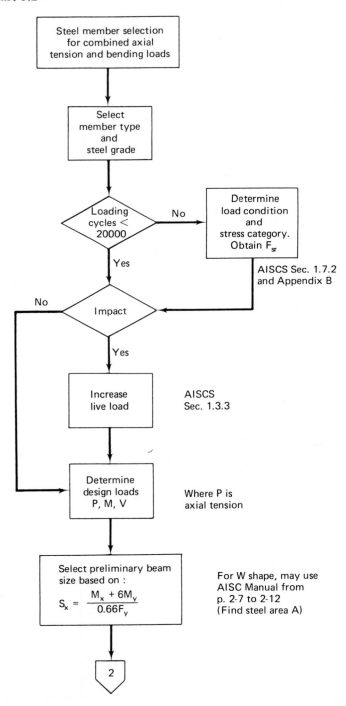

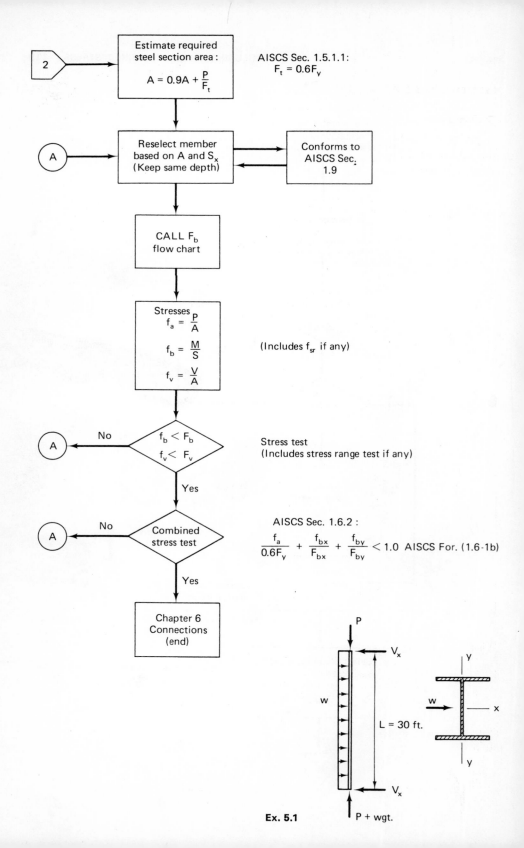

Ex. 5.1

5.6   ILLUSTRATIVE EXAMPLES

Example 5.1†

A 30-ft-long column is subjected to an axial compression load, $P = 600$ kips, and lateral uniform load, $w = 0.30$ kip/ft, that causes bending about the column weak axis as shown. Select an economical W shape to satisfy the AISCS. Use A36 steel.

Solution

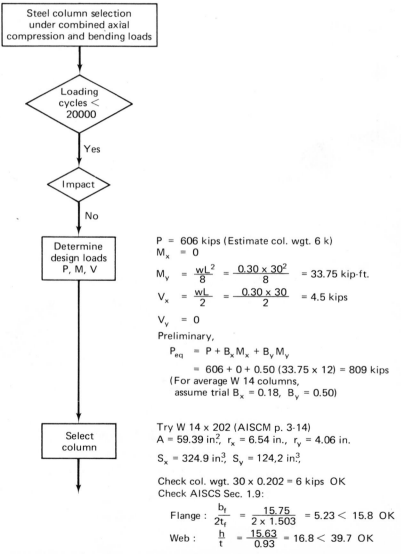

$P = 606$ kips (Estimate col. wgt. 6 k)

$M_x = 0$

$M_y = \dfrac{wL^2}{8} = \dfrac{0.30 \times 30^2}{8} = 33.75$ kip-ft.

$V_x = \dfrac{wL}{2} = \dfrac{0.30 \times 30}{2} = 4.5$ kips

$V_y = 0$

Preliminary,

$P_{eq} = P + B_x M_x + B_y M_y$

$= 606 + 0 + 0.50\,(33.75 \times 12) = 809$ kips

(For average W 14 columns, assume trial $B_x = 0.18$, $B_y = 0.50$)

Try W 14 x 202 (AISCM p. 3-14)

$A = 59.39$ in.², $r_x = 6.54$ in., $r_y = 4.06$ in.

$S_x = 324.9$ in.³, $S_y = 124.2$ in.³

Check col. wgt. $30 \times 0.202 = 6$ kips OK

Check AISCS Sec. 1.9:

Flange : $\dfrac{b_f}{2t_f} = \dfrac{15.75}{2 \times 1.503} = 5.23 < 15.8$ OK

Web : $\dfrac{h}{t} = \dfrac{15.63}{0.93} = 16.8 < 39.7$ OK

†An alternative solution of this problem will be found in Reference 1.7, page 183.

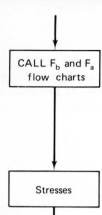

F_{by} = 27 ksi

(Since $\dfrac{b_f}{2t_f} = \dfrac{15.75}{2 \times 1.503} = 5.2 < 8.7$)

F_a = 14.34 ksi

(From AISCS Table 1-36,

for $\dfrac{K\ell}{r} = \dfrac{1.0\,(30 \times 12)}{4.06} = 88.7$)

$f_a = \dfrac{P}{A} = \dfrac{606}{59.39} = 10.2$ ksi

$f_{by} = \dfrac{M_y}{S_y} = \dfrac{33.75 \times 12}{124.4} = 3.26$ ksi

$f_{bx} = f_{vy} = 0$

$f_{vx} = \dfrac{V_x}{1.33A_f} = \dfrac{4.5}{1.33\,(15.75 \times 1.5)} = 0.095$ ksi

Stress test OK

$\dfrac{f_a}{F_a} = \dfrac{10.2}{14.34} = 0.711 > 0.15$

$\dfrac{f_a}{F_a} + \dfrac{C_{mx} f_{bx}}{(1 - f_a/F'_{ex}) F_{bx}} + \dfrac{C_{my} f_{by}}{(1 - f_a/F'_{ey}) F_{by}}$

$= 0.711 + 0 + \dfrac{1.0 \times 3.26}{(1 - 10.2/18.92)\,27} = 0.902 < 1.0$ OK

(F'_{ey} = 18.92 ksi obtain from AISCS Table 2,
p. 5-94, for $K\ell_b/r_b = 88.7$)

$\dfrac{f_a}{0.6F_y} + \dfrac{f_{bx}}{F_{bx}} + \dfrac{f_{by}}{F_{by}} = \dfrac{10.2}{0.6 \times 36} + 0 + \dfrac{3.26}{27} = 0.593 < 1.0$ OK

USE W 14 x 202

Remark:

Check capacity of W 14 x 193 by AISCM Modified
Interaction Formula and tables.
For W 14 x 193, $a_y = 139 \times 10^6$, $B_y = 0.481$

$\dfrac{K\ell}{r_y} = \dfrac{1.0\,(30 \times 12)}{4.05} = 88.9$, $F_a = 14.33$ ksi

Required tabular load — AISCM Modified for. (1.6-1a)

$= 606 + [0] + [0.481 \times 405 \times 1.0\, \dfrac{14.33}{27} \cdot \dfrac{139 \times 10^6}{139 \times 10^6 - 606 \times 360^2}]$

$= 840.7$ kips > 813 (from AISCM p. 3-14)
Hence, W 14 x 193 is *not* OK

Example 5.2

Same as Ex. 5.1, except the lateral uniform load causes bending about the column strong axis.

Solution

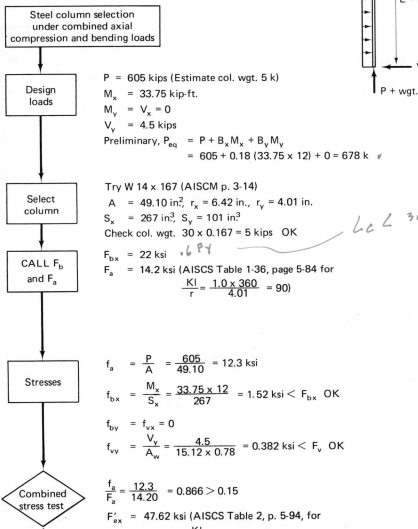

Steel column selection under combined axial compression and bending loads

Design loads

P = 605 kips (Estimate col. wgt. 5 k)
M_x = 33.75 kip-ft.
M_y = V_x = 0
V_y = 4.5 kips
Preliminary, P_{eq} = $P + B_x M_x + B_y M_y$
= 605 + 0.18 (33.75 × 12) + 0 = 678 k ✓

Select column

Try W 14 × 167 (AISCM p. 3-14)
A = 49.10 in^2, r_x = 6.42 in., r_y = 4.01 in.
S_x = 267 in^3, S_y = 101 in^3
Check col. wgt. 30 × 0.167 = 5 kips OK

$Lc < 30 < Lu$

CALL F_b and F_a

F_{bx} = 22 ksi   .6 F_y
F_a = 14.2 ksi (AISCS Table 1-36, page 5-84 for
$$\frac{Kl}{r} = \frac{1.0 \times 360}{4.01} = 90)$$

Stresses

$f_a = \dfrac{P}{A} = \dfrac{605}{49.10} = 12.3$ ksi

$f_{bx} = \dfrac{M_x}{S_x} = \dfrac{33.75 \times 12}{267} = 1.52$ ksi $< F_{bx}$ OK

$f_{by} = f_{vx} = 0$

$f_{vy} = \dfrac{V_y}{A_w} = \dfrac{4.5}{15.12 \times 0.78} = 0.382$ ksi $< F_v$ OK

Combined stress test

$\dfrac{f_a}{F_a} = \dfrac{12.3}{14.20} = 0.866 > 0.15$

F'_{ex} = 47.62 ksi (AISCS Table 2, p. 5-94, for
$$\frac{Kl_b}{r_b} = \frac{1.0 \times 360}{6.42} = 56)$$

AISCS For. (1.6-1a):

$$0.866 + \frac{1.0 \times 1.52}{(1 - 12.3/47.62)\,22} + 0 = 0.959 < 1.0 \text{ OK}$$

AISCS For. (1.6-1b) :

$$\frac{12.3}{0.6 \times 36} + \frac{1.52}{22} + 0 = 0.639 < 1.0 \text{ OK}$$

USE W 14 × 167

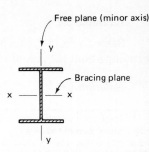

Free plane (minor axis)

Bracing plane

Example 5.3

Select an economical W shape for the same load conditions as in Ex. 5.2 but with adequate lateral supports provided along the column to cause the member to deflect in the plane of its minor axis.

Solution

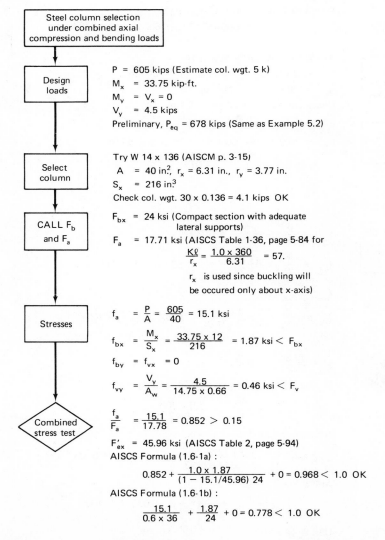

Steel column selection under combined axial compression and bending loads

Design loads

P = 605 kips (Estimate col. wgt. 5 k)
M_x = 33.75 kip-ft.
M_y = V_x = 0
V_y = 4.5 kips
Preliminary, P_{eq} = 678 kips (Same as Example 5.2)

Select column

Try W 14 x 136 (AISCM p. 3-15)
A = 40 in^2, r_x = 6.31 in., r_y = 3.77 in.
S_x = 216 in^3
Check col. wgt. 30 x 0.136 = 4.1 kips OK

CALL F_b and F_a

F_{bx} = 24 ksi (Compact section with adequate lateral supports)
F_a = 17.71 ksi (AISCS Table 1-36, page 5-84 for
$$\frac{K\ell}{r_x} = \frac{1.0 \times 360}{6.31} = 57.$$
r_x is used since buckling will be occured only about x-axis)

Stresses

$f_a = \dfrac{P}{A} = \dfrac{605}{40}$ = 15.1 ksi

$f_{bx} = \dfrac{M_x}{S_x} = \dfrac{33.75 \times 12}{216}$ = 1.87 ksi $<$ F_{bx}

f_{by} = f_{vx} = 0

$f_{vy} = \dfrac{V_y}{A_w} = \dfrac{4.5}{14.75 \times 0.66}$ = 0.46 ksi $<$ F_v

Combined stress test

$\dfrac{f_a}{F_a} = \dfrac{15.1}{17.78}$ = 0.852 $>$ 0.15

F'_{ex} = 45.96 ksi (AISCS Table 2, page 5-94)
AISCS Formula (1.6-1a) :
$$0.852 + \frac{1.0 \times 1.87}{(1 - 15.1/45.96)\,24} + 0 = 0.968 < 1.0 \ \text{OK}$$

AISCS Formula (1.6-1b) :
$$\frac{15.1}{0.6 \times 36} + \frac{1.87}{24} + 0 = 0.778 < 1.0 \ \text{OK}$$

USE W 14 x 136

Example 5.4

Select a column in a building frame for a 24-ft story height to support a 260-kip axial compression load and 100 and 120 kip-ft bending moments acting at the top and bottom ends, respectively, as shown. The column is braced against sidesway in the weak bending plane of the column (xz plane) but with possible sidesway in the strong bending plane (yz plane). Use AISCS and A36 steel.

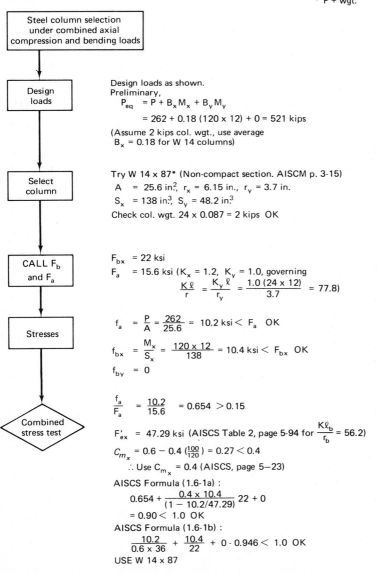

P = 260 k
M_{x_1} = 100 k-ft.
V_y

Braced plane against sidesway

L = 24 ft.

V_y
M_{x_2} = 120 k-ft.
P + wgt.

Solution

Steel column selection under combined axial compression and bending loads

Design loads

Design loads as shown.
Preliminary,
$$P_{eq} = P + B_x M_x + B_y M_y$$
$$= 262 + 0.18 (120 \times 12) + 0 = 521 \text{ kips}$$
(Assume 2 kips col. wgt., use average B_x = 0.18 for W 14 columns)

Select column

Try W 14 x 87* (Non-compact section. AISCM p. 3-15)
A = 25.6 in². r_x = 6.15 in., r_y = 3.7 in.
S_x = 138 in.³, S_y = 48.2 in³
Check col. wgt. 24 x 0.087 = 2 kips OK

CALL F_b and F_a

F_{bx} = 22 ksi
F_a = 15.6 ksi (K_x = 1.2, K_y = 1.0, governing
$$\frac{K\ell}{r} = \frac{K_y \ell}{r_y} = \frac{1.0 (24 \times 12)}{3.7} = 77.8)$$

Stresses

$$f_a = \frac{P}{A} = \frac{262}{25.6} = 10.2 \text{ ksi} < F_a \quad \text{OK}$$

$$f_{bx} = \frac{M_x}{S_x} = \frac{120 \times 12}{138} = 10.4 \text{ ksi} < F_{bx} \quad \text{OK}$$

$$f_{by} = 0$$

Combined stress test

$$\frac{f_a}{F_a} = \frac{10.2}{15.6} = 0.654 > 0.15$$

F'_{ex} = 47.29 ksi (AISCS Table 2, page 5-94 for $\frac{K\ell_b}{r_b}$ = 56.2)

C_{m_x} = 0.6 − 0.4 $(\frac{100}{120})$ = 0.27 < 0.4
∴ Use C_{m_x} = 0.4 (AISCS, page 5−23)

AISCS Formula (1.6-1a) :
$$0.654 + \frac{0.4 \times 10.4}{(1 - 10.2/47.29)} 22 + 0$$
$$= 0.90 < 1.0 \quad \text{OK}$$

AISCS Formula (1.6-1b) :
$$\frac{10.2}{0.6 \times 36} + \frac{10.4}{22} + 0 \cdot 0.946 < 1.0 \quad \text{OK}$$

USE W 14 x 87

* Remark: First trial was a W 14 x 111, but it was understressed by about 25%.

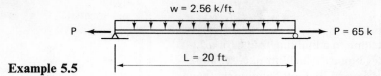

w = 2.56 k/ft.

P ← $\quad$ P = 65 k

L = 20 ft.

Example 5.5

A simply supported member has a span of 20 ft and carries a uniform live and dead load of 2.56 kips/ft (including its own weight), and axial *tension* load of 65 kips acting through the centroid of the member as shown. The compression flange of the member is laterally supported against local buckling. Select an economical W shape to satisfy the AISCS. Use A36 steel.

Solution

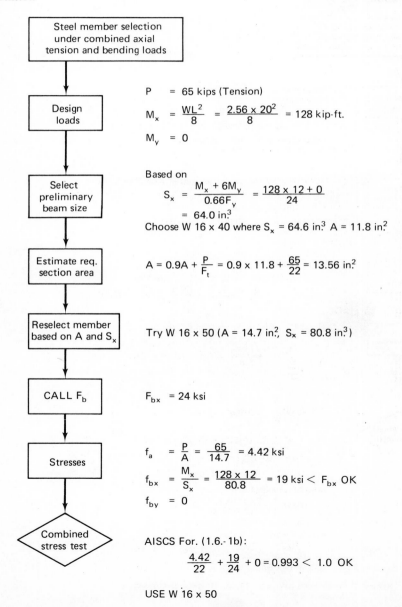

Steel member selection under combined axial tension and bending loads

↓

Design loads

$P = 65$ kips (Tension)

$M_x = \dfrac{WL^2}{8} = \dfrac{2.56 \times 20^2}{8} = 128$ kip-ft.

$M_y = 0$

↓

Select preliminary beam size

Based on
$$S_x = \frac{M_x + 6M_y}{0.66F_y} = \frac{128 \times 12 + 0}{24}$$
$$= 64.0 \text{ in.}^3$$
Choose W 16 × 40 where $S_x = 64.6$ in.^{3} $A = 11.8$ in.^{2}

↓

Estimate req. section area

$A = 0.9A + \dfrac{P}{F_t} = 0.9 \times 11.8 + \dfrac{65}{22} = 13.56$ in.^{2}

↓

Reselect member based on A and S_x

Try W 16 × 50 ($A = 14.7$ in.^{2}, $S_x = 80.8$ in.^{3})

↓

CALL F_b

$F_{bx} = 24$ ksi

↓

Stresses

$f_a = \dfrac{P}{A} = \dfrac{65}{14.7} = 4.42$ ksi

$f_{bx} = \dfrac{M_x}{S_x} = \dfrac{128 \times 12}{80.8} = 19$ ksi $< F_{bx}$ OK

$f_{by} = 0$

↓

Combined stress test

AISCS For. (1.6.-1b):
$$\frac{4.42}{22} + \frac{19}{24} + 0 = 0.993 < 1.0 \text{ OK}$$

USE W 16 × 50

PROBLEMS

5.1, 5.2, 5.3, 5.4, and 5.5. Rework Examples 5.1 through 5.5, changing yield point of steel from 36 to 50 ksi.

5.6. On the basis of allowable stress, ignoring column behavior and the effect of deflection on moment, select a stub cantilever for the conditions illustrated in Fig. 5.1, if the length is 3 ft, load P is 400 kips, with an eccentricity $e = 5$ in., using a W section of A36 steel bent about the strong axis. Check adequacy of selection by use of the AISCS interaction equation.

5.7. Redesign for the situation described in Problem 4.6, but with a lateral load of 250 lb/ft over the 36-ft column length.

5.8. A square box column, 40 ft in length, with sidewalls 24 in. apart (dimensioned as shown to the middle planes of the plates), carries an axial load of 600 kips and a lateral load of 400 lb/ft, acting normal to one of the flat sides. Determine the plate thickness to the nearest $\frac{1}{16}$ th in. if A36 steel is used.

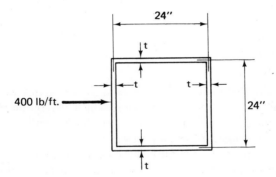

5.9. Redesign for the conditions of Problem 3.14 if the beam, in addition to the specified vertical and horizontal beam loads, carries a compressive force applied through the centroid at each end and amounting to 150 kips.

5.10. The beam in the figure is restrained laterally in the weak direction by ties as shown. For bending in the weak (horizontal) directions, it may be assumed as hinged at the center. Select an adequate member of A36 steel for this biaxial bending situation.

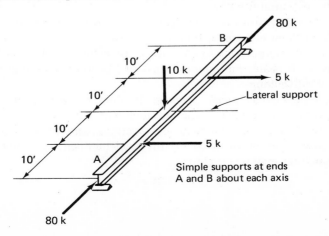

6

CONNECTIONS

6.1 INTRODUCTION

The component rolled plates and shapes of a steel structure are held together by means of fasteners (rivets or bolts) or by welds, which may either fuse the parts together into an integral unit or stitch them together intermittently as do fasteners. The fasteners and welds are used in the shop fabrication processes to build up members and again in the field erection to connect the separate members together to form the completed structure. If a member is too large to be shipped as a single unit, field connections must also be used to splice the member segments. Naturally, because of the relatively great cost of making field connections, such splices are kept to a minimum.

In truss construction, the tension and compression members meeting at a joint may be attached separately by fasteners to a *gusset plate* (see Fig.2.3), or, if welding is used, it may be possible to join them directly without the use of an auxiliary plate.

In building-frame construction the AISCS, Sec. 1.2, recognizes three basic types of beam-to-column framing:

Type 1, commonly designated as *rigid frame* or *continuous frame*, has beam-to-column or beam-to-girder connections that transmit the calculated moments and shears and have sufficient rigidity to provide the full continuity that has been assumed in the analysis. This means that no local kinks will be formed at allowable loads due to premature local yielding.

Type 2, designated as *simple framing*, provides flexible connections that are simply assumed to support the end reactions due to gravity load. The connections are purposely designed so as to permit beam end rotations relative to the column or girder to a degree that will permit one to ignore the incidental bending moments and small inelastic yielding that may be developed.

Type 3, *semirigid framing*, "assumes that the connections . . . possess a dependable and known moment capacity intermediate in degree between the rigidity of Type 1 and the flexibility of Type 2."

Type 1 moment-resisting connections are frequently used in the main building frames to resist wind and earthquake forces, with Type 2 connections being used in the remainder of the structure.

In addition to the material on connections in the AISCS and Commentary, the AISCM in Section 4 provides 142 pages of design information and tables that facilitate the proportioning of commonly used connections.

A connection is said to be loaded concentrically if the resultant (axial) force passes through the centroid of the fastener group or weld pattern. At low loads the distribution of stress transfer in such a connection is quite nonuniform, but prior to failure the local yielding of material tends to distribute the load uniformly to all parts of the connection. This fact, together with the results of experience and many laboratory tests to failure, permits the assumption in design that all parts of a concentrically loaded connection share equally in resisting the applied force. If the load does not pass through the centroid of the connection, or if the connection must transmit moment as well as axial force, the force system may be reduced to a force and a couple at the connection centroid. Each element of the connection is assumed to resist the axial component of force uniformly and to resist the moment in proportion to the distance of that element from the centroid of the connection. Design rules based on the foregoing assumptions have been found to be safe and workable.

Design recommendations and allowable stresses in connections as supplied in the AISCS are very largely backed up and based on recommendations of the Research Council on Riveted and Bolted Joints and the American Welding Society, for bolted and welded connections, respectively.

Each of the four general methods of making structural connections (rivets, bolts, pins, and welds) will be discussed and simple design examples presented. Direct use, by reference, will be made of the great amount of information available on this subject in the AISCM.

The overall strength and safety of a structure may be directly dependent on the connections that join main members. Such connections should be shown explicitly and in detail on the design drawings, both in the interest of safe design and for economy in view of bid-price allowances that might otherwise be made for contingencies.

6.2 RIVETED AND BOLTED CONNECTIONS

For many years riveting was the accepted method for making connections. However, the use of rivets has declined rapidly due to the development and economic advantages of welding and high-strength bolts. The advent of both welding and high-strength bolting has made possible the advantageous use of a combination of these fastening methods, such as shop fabrication by welding followed by the use of high-strength bolts for the field connections. In this way the advantages of each procedure are realized, as the welding is under shop-controlled conditions with the members positioned to produce good welds and economy in fabrication. The advantages of rapid assembly while the members are being held in position in the field are obtained through the use of high-strength bolts.

Rivets and bolts, as shown in Fig. 6.1, transmit force from one plate element to another by either single shear or double shear. In more complex joints with multiple plates interacting, more than two shear planes may occur.

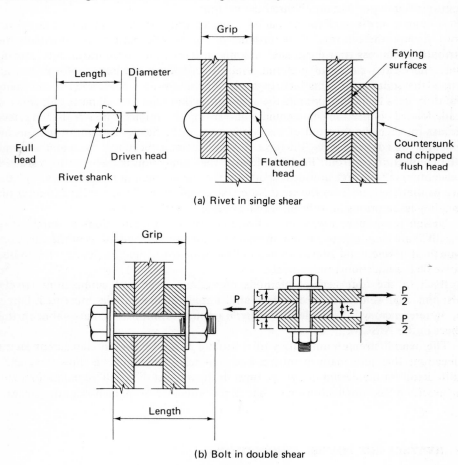

(a) Rivet in single shear

(b) Bolt in double shear

Fig. 6.1 Rivets and bolts.

Rivets are manufactured with a special full (acorn) head and are installed in holes that are punched or drilled $\frac{1}{16}$ in. larger in diameter than the rivets. The rivets are usually heated to approximately 1800°F before placing in the hole, and a second head is formed by means of a riveting hammer or a machine pressure type of riveter.

During the process of forming this head, the shank is upset so that it now completely fills the hole (see Fig. 6.1). The formed head can be either a full button head, a flattened head, or a countersunk and chipped head. The chipping is usually done with a chisel and air hammer after the rivet has cooled. Rivets with countersunk and chipped heads have less strength and cost considerably more than those with full or flattened heads. During cooling the rivet shrinks, setting up tensile forces in the shank, which may approach the yield point of the material. This residual tension force is unpredictable and may vary from practically nothing to a stress equal to the yield point. Because of this uncertainty, the clamping force exerted on the joined material is neglected in design calculations.

Rivets shall conform to the provisions of the "Specifications for Structural Rivets," ASTM A502, Grades 1 or 2. The size of rivets used in ordinary steel construction ranges from $\frac{5}{8}$ to $1\frac{1}{2}$ in. in diameter by $\frac{1}{8}$-in. increments. The allowable stresses are tabulated in the AISCM, pages 4-3 to 4-11 inclusive, in accordance with AISCS, Sec. 1.5.2, page 5-20. The AISCM, pages 4-112 to 4-121, should here be studied with regard to detailing practice, erection clearances, recommended spacing, and conventional signs for use on drawings for both rivets and bolts.

Unfinished bolts, also known as ordinary or common bolts, shall conform to the "Specifications for Low Carbon Steel Externally and Internally Threaded Standard Fasteners," ASTM A307. These bolts range from $\frac{5}{8}$ to $1\frac{1}{2}$ in. in diameter by $\frac{1}{8}$-in. increments. The allowable stresses are also tabulated in the AISCM, pages 4-3 to 4-11 inclusive, in accordance with AISCS, Sec. 1.5.2. Because of the uncertainty as to whether or not the threaded portion of unfinished bolts extends into the shear plane, their allowable stresses are considerably less than those permitted for rivets or high-strength bolts. Their use is usually restricted to structures subjected to static loads, and for secondary members such as purlins, girts, and bracing. AISCS Sec. 1.15.12, pages 5-41 and 5-42, lists specific connection types for which A307 bolts may *not* be used.

Since the initial tension developed by rivets and unfinished bolts is uncertain and possibly very small, no frictional resistance on the faying surfaces is assumed, and slip may occur at low shearing loads. This brings the rivets or bolts into bearing, and the mode of stress transfer is as shown in Fig. 6.2.

Fig. 6.2   Stress transfer by shear and bearing in a riveted or bolted bearing type connection.

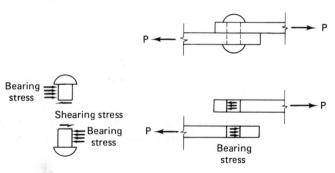

The bearing stress f_p on the contact area between the fasteners and the connected plates is defined as the transmitted shear load P divided by the total effective bearing area, dt, where d is the nominal diameter of the rivet or unfinished bolt, and t the thickness of the connected material.

The shearing stress f_v in fasteners is defined as the transmitted shear load P divided by the total effective shearing area A_v:

$$f_v = \frac{P}{A_v}$$

where

$$A_v = \frac{\pi d^2}{4} \qquad \text{for single shear}$$

$$= 2\frac{\pi d^2}{4} \qquad \text{for double shear}$$

Rivets and unfinished bolts are both acceptable in tension-type connections (such as hangers and moment connections), and the allowable load per fastener is tabulated on page 4-3 of the AISCM. Loads for rivets are based on the gross cross-sectional area using the nominal diameter, and for bolts on the tensile stress area, which is defined as

$$0.7854\left(D - \frac{0.9743}{n}\right)^2 \qquad \text{(AISCS, Sec. 1.5.2.1)}$$

where
$$D = \text{bolt diameter}$$
$$n = \text{number of threads per inch}$$

High-strength bolts, a relatively new type of fastener, are torqued to a high tensile stress in the shank, thus developing a dependable clamping pressure. In this case, shearing stress is transferred by friction at working loads, as illustrated in Fig. 6.3. High-strength bolts have found rapid adoption as the preferred fastener for field connections, resistance to stress reversal, impact loads, and other applications where joint slip is not desirable. Ease of installation is another desirable attribute.

High-strength bolts are available in two different strength levels, and are used in accordance with the provisions of the "Specification for Structural Joints Using ASTM A325 or A490 Bolts" as approved by the Research Council on Riveted and Bolted Structural Joints of the Engineering Foundation, April 18, 1972. A449 bolts have the same chemistry as A325 bolts but may have a smaller head size and are threaded full length.

In the friction-type connection, the bolts are not actually stressed in shear and are not in bearing, since no slip occurs at allowable loads. However, a shear stress is specified as a matter of convenience, and the number of fasteners is determined in the same manner as for other riveted or bolted connections. High-strength bolts are tightened so as to produce a minimum initial tension in the bolt shank equal to the proof load, or approximately 70 per cent of the tensile strength of the bolt. In order to obtain the specified initial tension in the bolts, they are usually tightened with properly calibrated wrenches or by the turn-of-nut method. A third new method of

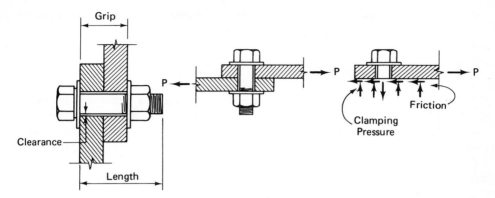

Fig. 6.3   Friction-type connection.

installation was added to the Bolt Specification in 1972, whereby a direct tension indicator is utilized. See the first revised (1972) printing of AISCM, page 5-197. Specifications require that hardened washers be used under the turned element when using the calibrated-wrench method and for A490 bolts when using the turn-of-nut method. Hardened washers are required under both head and nut when using A490 bolts to connect material with a yield point of less than 40 ksi. Beveled washers are required when an outer face of the connection has a slope greater than 1:20. In the turn-of-nut method, nuts are first brought to a "snug" fit, defined as the condition when all surfaces are in good contact and no free rotation of the nut is possible. Bolts under 8 diameters or 8 in. in length are then given an additional one half turn of the nut; over 8 diameters or 8 in., two thirds of a turn; and if both faces have a 1:20 slope, without use of washers, the nuts are to be given a three quarter turn, regardless of length.

Resistance to slip is determined by the amount of bolt tension and the condition of the contact surfaces in a given connection. Connections having painted contact surfaces or contact surfaces of unrusted mill scale offer the least resistance to slip; rusted surfaces that have been well cleaned may provide up to two times as much resistance.

Rivets and bolts in combined shear and tension are proportioned according to AISCS, Sec. 1.6.3. In the case of rivets and bearing-type bolted connections, the allowable tensile stresses are reduced if the simultaneous shear stress exceeds a certain value. For example, the allowable tensile stress for A502 Grade 1 rivets is

$$F_t = 28.0 - 1.6f_v \le 20.0 \text{ ksi}$$

This says, in effect, that the allowable tensile stress is in no case greater than 20 ksi, and if the shear stress exceeds 5 ksi, allowable tensile stress will be less than 20 ksi, as given by the formula. Similar formulas for Grade 2 rivets, and for unfinished and high-tensile bolts in bearing-type joints, are provided in Sec. 1.6.3 of the AISCS.

In the case of friction-type joints with high-tensile bolts, the allowable shear stress is dependent on the existence of adequate initial tension. Hence, if additional tension is applied, the clamping force is reduced and AISCS, Sec. 1.6.3, requires:

for A325 and A449 bolts $\quad F_v = 15.0\left(1 - \dfrac{f_t A_b}{T_b}\right)$

for A490 bolts $\quad F_v = 20.0\left(1 - \dfrac{f_t A_b}{T_b}\right)$

In these formulas, "f_t is the average tensile stress due to a direct load applied to all of the bolts in a connection and T_b is the specified pretension load of the bolt."

At this point the reader should review in detail the AISCS and AISCM information on rivets and bolts, as follows: AISCS, Sec. 1.5.2.1 and 1.5.2.2, together with Table 1.5.2.1, which lists allowable stresses in tension and shear for rivets and bolts, including friction-type connections. Also read Sec. 1.16, page 5-42 of AISCS, which provides detailed information on minimum pitch, minimum edge distance, and so forth, and also Secs. 1.23.4 and 1.23.5, page 5-51, which define good fabrication practice for riveted and bolted joints. Finally, note the detailed dimension and weight information regarding rivets and bolts on pages 4-116 to 4-121 of AISCM, and the complete tabulation of rivet and bolt strength values, at prescribed allowable stresses, on pages 4-3 to 4-11. A reason for the textual omission of much of the information just referred to is the desirability of fostering the direct use of specification and manual information in the handling of design problems.

Examples 6.1, 6.2, and 6.3 now illustrate the design of simple concentrically loaded riveted and bolted joints.

Example 6.1 *Lap Joints*

A lap joint is simply a joint in which two members overlap and are connected to each other with some type of fastener. The lap joint is not a desirable structural connection and should be used only for minor connections. The eccentricity of the loads causes secondary bending stresses in the members and at least two fasteners should be used on each line. (Edge distance, net section, and spacing checks omitted. See Ex. 6.2.)

A. Two $\frac{1}{4} \times 8$-in. A36 plates are to be connected using $\frac{3}{4}$-in. A502 Grade 1 rivets, as shown in the figure. How many rivets are required to transfer the load $P = 30$ kips?

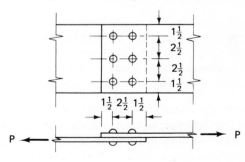

Solution

Rivets are critical in single shear or bearing (refer to AISCS, Sec. 1.5.2.1):

bearing value = $36 \times 1.35 \times \frac{3}{4} \times \frac{1}{4} = 9.1$ kips/rivet $\quad (F_p = 1.35F_y)$

$$\text{shear value} = 15 \times 0.442 = 6.63 \text{ kips/rivet} \qquad (F_v = 15 \text{ ksi})$$

Single shear governs:

$$\text{no. rivets reqd} = \frac{30}{6.63} = 4.5$$

Use 6 rivets for symmetry. (Symmetry in the placement of fasteners is desirable in order to avoid secondary stresses that add to the already complex stress distribution. Note that rivet values could have been obtained directly from AISCM, pages 4-4 and 4-5.)

B. Same as (A) except use $\frac{3}{4}$-in.-diameter A325 bolts in a friction-type connection and increase the load P to 30 kips. (Bearing is not restricted—AISCS, Sec. 1.5.2.2.)

Solution

$$\text{shear value} = F_v A_b = 15 \times 0.442 = 6.63 \text{ kips/bolt}$$

(Or refer to AISCM, page 4-4.)

$$\text{no. bolts reqd} = \frac{30}{6.63} = 4.5$$

Use 6 bolts.

C. Same as (B) except in bearing-type connection with threads excluded from shear plane.

Solution

$$\text{bearing on plate} = 36 \times 1.35 \times \tfrac{3}{4} \times \tfrac{1}{4} = 9.1 \text{ kips/bolt} \qquad \text{(governs)}$$
$$\text{shear value} = 22 \times 0.442 = 9.72 \text{ kips/bolt} \qquad (F_v = 22 \text{ ksi})$$
$$\text{no. bolts reqd} = \frac{30}{9.1} = 3.3$$

Use 4 bolts.

Example 6.2   *Butt Splice*

A. A butt splice is to be designed as shown in the figure. Use $\frac{3}{4}$-in.-diameter A325 bolts in a friction-type connection and A36 steel. $P = 80$ kips. Determine the number of bolts required on each side of the splice. Bolts are in double shear.

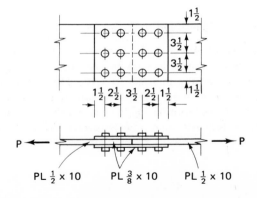

PL $\frac{1}{2}$ × 10   PL $\frac{3}{8}$ × 10   PL $\frac{1}{2}$ × 10

Solution

capacity of one bolt in double shear $= 2F_vA_b = 2 \times 15 \times 0.442$

$$= 13.25 \text{ kips} \quad \text{(Or see AISCM, page 4-4.)}$$

$$\text{no. bolts reqd} = \frac{80}{13.25} = 6.05$$

Use 6 bolts (3 bolts per line).

B. Same as (A) except use $\frac{3}{4}$-in.-diameter A325 bolts in a bearing-type connection with threads excluded from the shear plane. Increase P to 90 kips.

Solution

capacity of one bolt in double shear $= 2 \times 0.442 \times 22 = 19.45 \text{ kips}$

capacity of one bolt in bearing on $\frac{1}{2}$-in. plate

$$= 36 \times 1.35 \times 0.75 \times \tfrac{1}{2} = 18.2 \text{ kips} \quad \text{(governs)}$$

$$\text{no. bolts reqd} = \frac{90}{18.2} = 4.95$$

Use 6 bolts in two rows of 3 bolts each. Check plate net section capacity and bolt clearance for Ex. 6.2(B) [also applicable to Ex. 6.2(A)].

For spacing as shown in the figure for Ex. 6.2(A), assume sheared edges and refer to Table 1.16.5, AISCS, page 5-43:

$$\text{minimum edge distance} = 1\tfrac{1}{4} \text{ in.} < 1\tfrac{1}{2} \quad \text{OK}$$
$$\text{preferred minimum pitch} = 3 \times \tfrac{3}{4} = 2\tfrac{1}{4} < 3\tfrac{1}{2} \quad \text{OK}$$

Net section capacity:

$$[10 - 3(\tfrac{3}{4} + \tfrac{1}{8})](\tfrac{1}{2})(22) = 81.1 \text{ kips} < 90 \qquad \text{OK for } P = 80 \text{ kips in Ex. 6.2(A), but } NG \text{ for Ex. 6.2(B)}$$

Change to $\frac{9}{16} \times 10$ plate

$$[10 - 3(\tfrac{3}{4} + \tfrac{1}{8})](\tfrac{9}{16})(22) = 91.3 \text{ kips} > 90 \text{ kips} \qquad \text{OK}$$

Example 6.3 *Bracket Connection*

A tension member, two L $4 \times 3 \times \frac{1}{2}$, carries a tension load P of 120 kips with a direction of 30° from the horizontal axis. A bracket utilizing a structural tee section will be used to connect the tension member by rivets and will be joined to a column flange by high-strength bolts with a friction-type connection, as shown in the figure. Determine the required size and number of rivets and bolts. Use A502 Grade 2 rivets, A490 high-strength bolts, and A36 steel in accordance with the AISCS.

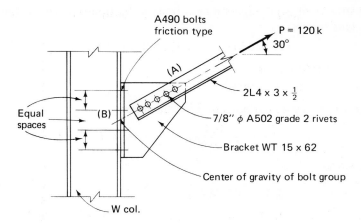

Solution

A. Connection A—tension angles to bracket (try WT 12×65, $t_w = 0.565$ in.).
Allowable shear stress:

$$F_v = 20 \text{ ksi} \qquad \text{(A502 Grade 2, AISCS, Table 1.5.2.1)}$$

Allowable bearing stress:

$$F_p = 48.6 \text{ ksi} \qquad \text{(A36 steel, AISCS, Sec. 1.5.2, Appendix A, page 5-72.)}$$

Try $\frac{7}{8}$-in.-diameter rivets ($A_b = 0.601$ in.^{2}).
Double shear strength:

$$2F_v A_b = 2 \times 20 \times 0.601 = 24 \text{ kips/rivet} \qquad \text{(governs)}$$

Bearing strength on 0.565 tee web equals $(48.6)(0.565)(\frac{7}{8}) = 24.1$ kips/rivet.
Check net section capacity of double-angle member. Refer to AISCM, page 1-60,
or 1-76.

$$\text{net area} = 6.50 - (2)(0.5)(\tfrac{7}{8} + \tfrac{1}{8}) = 6.50 - 1.0 = 5.5 \text{ in.}^2$$

$$\text{tensile capacity} = (5.5)(22) = 121.0 > 120 \text{ kips} \qquad \text{OK}$$

$$\text{no. rivets reqd} = \tfrac{120}{24} = 5$$

Arrange as shown, at $2\frac{1}{2}$ in. center to center. Detailer to check layout for clearances
and edge distances.

B. Connection B—bracket to column flange. Assume the tension load passes through
the center of gravity of the bolts. Then the shear and tension components of the
tension are
Tension component:

$$T = P \cos 30° = 120 \times 0.866 = 104 \text{ kips}$$

Shear component:

$$V = P \sin 30° = 120 \times 0.5 = 60 \text{ kips}$$

Allowable stresses of A490 high-strength bolts in friction-type connection are
Allowable tensile stress:

$$F_t = 54 \text{ ksi} \qquad \text{(AISCS Table 1.5.2.1)}$$

Allowable shear stress:

$$F_v = 20 \text{ ksi}$$

Bolt tensile stress—assume 8 bolts of $\frac{7}{8}$ in. diameter:

$$f_t = \frac{T}{nA_b} = \frac{104}{8 \times 0.601} = 21.6 \text{ ksi} < F_t \qquad \text{OK}$$

Bolt shear stress:

$$f_v = \frac{V}{nA_b} = \frac{60}{8 \times 0.601} = 12.5 \text{ ksi}$$

Pretension of $\frac{7}{8}$-in. bolt, $T_b = 49$ kips (Table 1.23.5, page 5-52, AISCS). Reduced
allowable shear stress due to combined tension is

$$F_v = 20\left(1 - \frac{f_t A_b}{T_b}\right) \qquad \text{(see text and AISCS, Sec. 1.6.3)}$$

$$= 20\left(1 - 21.6 \times \frac{0.601}{49}\right) = 14.7 \text{ ksi} > f_v \qquad \text{OK}$$

Use 8 bolts of $\frac{7}{8}$ in. diameter.

Note that prying force on A490 bolts should also be investigated. The calculations
involved will be demonstrated in a later example.

6.3 PINNED CONNECTIONS

Pinned connections are sometimes used in bridge-bearing supports with the
purpose of permitting end rotation. They are also used to connect pin-connected
members of the types discussed in Chapter 2. Ranging from 2 to 10 in. or more in
diameter, they are designed in a manner analogous to bearing connections of bolts,
but with lower allowable stresses and with the added requirement to check stress due
to bending in the pin itself. Details of standard pins and caps or nuts to hold them in
position are provided on page 4-129 of the AISCM. Allowable stresses due to shear,
bending, and in bearing are $0.4F_y$, $0.75F_y$, $0.9F_y$, respectively, as stated in AISCS,
Secs. 1.5.1.2, 1.5.1.4.3, and 1.5.1.5.1, respectively. Although the actual distribution
of stress in a short circular beam is complex, designs have been found to be satisfactory
when based on simple beam theory and on average stress due to shear and bearing.
Bending moments may be conservatively calculated on the assumption that forces
are concentrated at the centers of bearing areas.

On the basis of assumed locations of acting forces, bending moments and shears
may be determined, and a preliminary selection of required pin diameters may then be
determined on the basis of whichever stress (bending or shear) is critical. The bearing

stress may then be checked, and revision of either the pin diameter or length of bearing may then be made if required.

In bending

$$f_b = \frac{M}{S} = \frac{32M}{\pi d^3}$$

Hence, to maintain $f_b = F_b$, the required pin diameter is

$$d_{req} \geq \sqrt[3]{\frac{32M}{\pi F_b}} \qquad (6.1)$$

In shear

$$f_v(\text{avg}) = \frac{4V}{\pi d^2}$$

and, similarly, the required pin diameter for shear is

$$d_{req} \geq \sqrt{\frac{4V}{\pi F_v}} \qquad (6.2)$$

Example 6.4 *Pin Connection*

Although the connection shown is of a type not commonly used in recent years, it illustrates very adequately the essential problems in pin design. For A36 steel the allowable stresses are, by reference to the AISCS sections referred to in the previous paragraph,

for shear, $F_v = 0.4F_y = 14.5$ ksi

for bending, $F_b = 0.75F_y = 27.0$ ksi

for bearing, $F_p = 0.90F_y = 33.0$ ksi

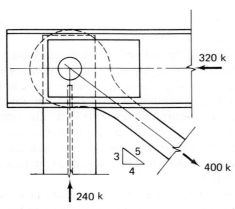

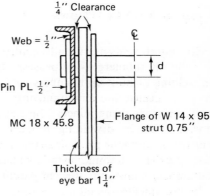

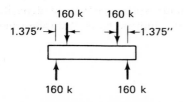

Horizontal components

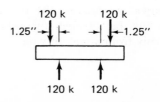

Vertical components

Bending moment:

$$M_h = 160 \times 1.375 = 220 \text{ kip-in.}$$
$$M_v = 120 \times 1.250 = 150 \text{ kip-in.}$$
$$M = \sqrt{220^2 + 150^2} = 266 \text{ kip-in.}$$

Pin diameter required by moment [Eq. (6.1)]:

$$d = \sqrt[3]{\frac{32M}{\pi F_b}} = \sqrt[3]{\frac{32 \times 266}{\pi \times 27}} = 4.65 \text{ in.}$$

Pin diameter required by shear [Eq. (6.2)]:

$$\text{shear load } V = \tfrac{400}{2} = 200 \text{ kips}$$

$$d = \sqrt{\frac{4V}{\pi F_v}} = \sqrt{\frac{4 \times 200}{\pi \times 14.5}} = 4.2 \text{ in.}$$

Pin diameter required by bearing:

$$d = \frac{P}{tF_p} = \frac{120}{0.75 \times 33} = 4.85 \text{ in.} \qquad \text{(flange of vertical strut)}$$

$$d = \frac{160}{33 \times 1.02} = 4.75 \text{ in.} \qquad \text{(on chord)}$$

$$d = \frac{200}{33 \times 1.25} = 4.85 \text{ in.} \qquad \text{(on eyebar)}$$

Use pin diameter $d = 5.0$ in.

[*Note:* Attachment of pin plate to channel web is an assigned problem; No. 6.3.]

6.4 WELDED CONNECTIONS

Structural welds are usually made either by the manual shielded-metal-arc process or by the submerged-arc process, the latter being especially suited to automatic shop welding of built-up members with controlled positioning. In either process the heat of an electric arc simultaneously melts the welding electrode and the adjacent steel in the parts being joined. The electrode is deposited in the weld as filler metal. The wide adoption of welding in recent years has required improved control of steel chemistry in order to provide steels that are "weldable," that is, steels that can be joined together with sound metal, of adequate strength and ductility, and with minimal metallurgical damage to adjacent parent metal.

In the shielded-metal-arc process, pictured in Fig. 6.4, the electrode coating creates a gaseous shield that protects the molten weld metal from the atmosphere. In the submerged-arc process, the arc occurs underneath a previously deposited fusible powdered flux that blankets the welding zone, and the bare electrode usually is fed automatically from a reel of wire.

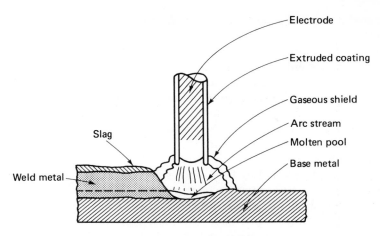

Fig. 6.4   Shielded-metal-arc welding process.

The adoption of rules governing the qualification of welders and welding processes together with quality control of all materials have brought welding to the point where it is today permitted for practically all steel fabrication for both shop and field connections. Welding offers many advantages, which may be briefly outlined as follows:

1. Simplicity of design details, efficiency, and minimum weight are achieved because welding provides the most direct transfer of stress from one member to another.

2. Fabrication costs are reduced because fewer parts are handled and operations such as punching, reaming, and drilling are eliminated.

3. There is a saving in weight in main tension members since there is no reduction in area due to rivet and bolt holes. Additional saving is also achieved because of the fewer connecting parts required.

4. Welding provides the only plate-joining procedure that is inherently air- and watertight and hence is ideal for water and oil storage tanks, ships, and so forth.

5. Welding permits the use of fluidly changing lines that enhance the structural and architectural appearance, as well as reduce stress concentrations due to local discontinuities.

6. Simple fabrication becomes practicable for those joints in which a member is joined to a curved or sloping surface, such as structural pipe connections.

7. Welding simplifies the strengthening and repair of existing riveted or welded structures.

The two most common types of welds are fillet welds and groove welds. Fillet welds are used to attach a plate to another plate or member in either a parallel (lapped) or protruding (tee) position, as shown in Figs. 6.5(a). Groove welds, as shown in Fig. 6.5(b), retain the continuity of plate elements that are butt joined along their edges. Groove welds require special edge preparation and careful fit up, and when welded from both sides, or from one side with a backup strip on the far side, they may be said to achieve *complete penetration* and may be stressed as much as the weakest piece that has been joined. Incomplete-penetration groove welds are used only when the plates are not required to be fully stressed and full continuity is not required. Complete-penetration groove welds are also used for corner or tee joints when full plate development is required.

The AISCM, pages 4-133 to 4-141, shows in cross section many types of groove welds for various plate arrangements, edge preparation as required, whether manual shielded-metal-arc or submerged-arc process is used, and whether the welds are to be complete or incomplete penetration. All these types are as permitted by AISCS, Sec. 1.17.2, "Qualification of Weld and Joint Details," as sanctioned by the AWS (American Welding Society). AISCS, Table 1.17.2, lists the types of electrodes (according to base metal, strength, coating, and so forth) that may be used for groove welds. For full effectiveness, the weld metal must in all cases have equivalent or greater strength than that of the plates that are joined.

AISCS, Table 1.5.3, page 5-21, indicates that allowable stresses in compression,

Fig. 6.5   Types of welded joints.

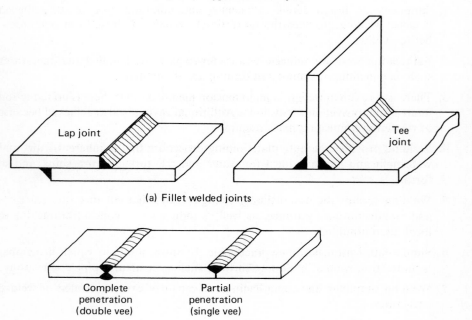

(a) Fillet welded joints

(b) Groove welded butt joints

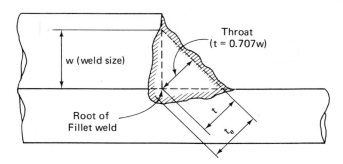

Fig. 6.6   Fillet weld nomenclature.

tension, and shear for full-penetration groove welds are the same as for the base metal. Partial-penetration groove welds are permitted full stress effectiveness only for compression normal to the effective throat.[†] They may be stressed in tension on the effective throat at a reduced allowable stress equal to that allowed in shear.

Standard weld symbols to be used in steel detailing are shown on page 4-132 of the AISCM. Their correct use is illustrated in many sketch details shown in the AISCM, pages 4-28 to 4-110, and familiarity with this usage can be gained by practice and by reference to these sketches.

Fillet welds are more easily made than groove welds because of the greater fit-up tolerances allowed. As shown in Fig. 6.5, fillet welds are commonly used in connecting lapped plates or projecting plate elements to another plate or member.

The allowable force transmitted by a unit length of fillet weld is equal to the product of the *effective throat dimension* multiplied by the allowable *shear* stress, as listed in Table 1.5.3, for the electrode and base metal specified therein. The effective throat dimension is illustrated in Fig. 6.6, where w is the nominal weld size and t the throat dimension. When the welded faces of the joined parts are at 90°, as shown,

$$t = 0.707w$$

When the manual shielded-metal-arc process is used, the effective throat is equal to the dimension t. When the submerged-metal-arc process is used, greater heat input produces a deeper penetration and an effectively greater throat dimension is allowed. Thus, for the submerged-arc process, AISCS, Sec. 1.14.7, permits the effective throat, T_e, to be taken as equal to the weld size, w, when w is $\frac{3}{8}$ in. or less, and equal to $t +$ 0.11 when w is greater than $\frac{3}{8}$ in.

Table 1.5.3 lists allowable stresses for various electrode and base-metal combinations for fillet welds. These allowables are stated to be for "shear stress on effective throat of fillet weld *regardless of direction of application of load.*" What is really meant here is that the *resultant stress* on the effective throat is taken to be *equivalent* to shear stress in determining the allowable force on a unit length of weld. This concept is illustrated in Fig. 6.7, where the z axis is arbitrarily shown along the fillet weld throat

[†]See AISCM, pages 4-139 to 4-141, for a definition of the effective throat (T_e) of various partial-penetration groove welds.

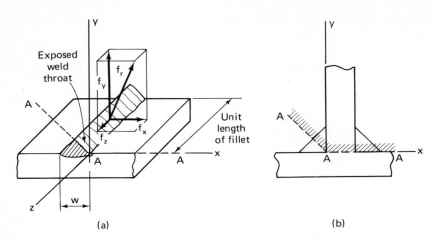

Fig. 6.7 Stress components and resultant stress on fillet weld throat.

and the x and y axes are in the surface planes of the tee joint, as shown in Fig. 6.7(b). The region above lines A-A-A has been removed in the pictorial representation of Fig. 6.7(a) in which the section through the fillet throat is exposed. The resultant stress f_r, having components f_x, f_y, and f_z, must be kept below the permissible stress values listed in AISCS, Table 1.5.3. It may be noted that f_z is a pure shear component, directed along the weld, and that f_x and f_y include both shear and normal components. The stress resultant, f_r, is

$$f_r = \sqrt{f_x^2 + f_y^2 + f_z^2} \qquad (6.3)$$

The x, y, and z axes may be oriented in any arbitrary direction. The use of the stress resultant as a strength criterion has been shown to be adequate by many tests of welded connections, including those that are eccentrically loaded.

Fillet welds are specified on drawings and in design calculations by their size, w, which vary from $\frac{3}{16}$ to $\frac{1}{2}$ in. in $\frac{1}{16}$-in. increments, and which vary in $\frac{1}{8}$-in. increments for sizes greater than $\frac{1}{2}$ in. Weld sizes of $\frac{3}{16}$, $\frac{1}{4}$, and $\frac{5}{16}$ in. are favored because they can be made by a single pass of the electrode. The amount of filler metal increases as the square of the weld size. Thus time and cost of welding increase disproportionately as the weld size increases.

There are other limitations on fillet-weld size. A small weld at the edge of a thick plate is chilled rapidly, which has an embrittling effect, and it may crack as it tries to shrink when it cools while being restrained by the heavy plate. Minimum permitted sizes of fillet welds, in reference to the thickest plate joined, are listed in Table 1.17.5, page 5-44, of the AISCS. Welds along the edges of plates thicker than $\frac{1}{4}$ in. are also limited as to *maximum* size, which can be no greater than the plate thickness minus $\frac{1}{16}$ in. (AISCS, Sec. 1.17.6). When a fillet weld is terminated, small sections near the ends are not fully effective. Rules of design covering this and other related topics are covered in AISCS, Secs. 1.17.7. to 1.17.10.

Allowable weld strength values per lineal inch are most conveniently quoted in terms of the weld size, w, even though determined by the resultant stress on the throat. Weld sizes are called for in multiple values of $\frac{1}{16}$ in. It is therefore convenient in design to use the strength value of a $\frac{1}{16}$-in. weld as a basic unit. Thus, if F_r is the allowable stress from AISCS, Table 1.5.3, the allowable strength value per inch of weld length and per $\frac{1}{16}$ in. of fillet weld size will be termed $\bar{q}$, where

$$\bar{q} = \tfrac{1}{16}(0.707)F_r$$

Letting D = the number of $\frac{1}{16}$ in. in a weld (for example, $D = 4$ for a $\frac{1}{4}$ in. weld), the allowable design force per inch of fillet weld is termed q where

$$q_a = \bar{q}D$$

For submerged-arc welds, the greater effective throat was previously defined (AISCS, Sec. 1.14.7). Values per $\frac{1}{16}$ in. of weld size are equal to $F_r/16$ for welds of $\frac{3}{8}$ in. or less. For weld sizes over $\frac{3}{8}$ in. the force values are the same as for the metal-arc process, plus a fixed bonus of $0.11F_r$. These values are listed and explained in Table 6.1 for the six different allowable stress levels established in AISCS, Table 1.5.3.

Table 6.1

Permissible force transfer of fillet welds, $\bar{q}$, per linear inch and per $\frac{1}{16}$ in. of weld size, in kips per inch, for various permissible stress levels from AISCS, Table 1.5.3.

		18.0	21.0	24.0	27.0	30.0	33.0
Permissible shear stress on effective throat of fillet weld (ksi)		18.0	21.0	24.0	27.0	30.0	33.0
Metal-arc electrode		E60	E70	E80	E90	E100	E110
Allowable weld force, kips per in. of length, per $\frac{1}{16}$ in. of weld size	Metal-arc process	0.80	0.93	1.06	1.19	1.33	1.46
	Submerged-arc, size $\frac{3}{8}$ in. or less	1.12	1.31	1.50	1.69	1.88	2.06
Submerged-arc/metal-arc kips per inch bonus, size over $\frac{3}{8}$ in.		1.98[a]	2.32[a]	2.68[a]	2.97[a]	3.30[a]	3.63[a]

[a]For submerged-arc welds over $\frac{3}{8}$ in. in size, use metal-arc process values from Table 6.1 and add bonus. For example, for a $\frac{1}{2}$-in. submerged-arc weld, $\frac{1}{2} = \frac{8}{16}$, and for Grade 80 submerged-arc weld, allow $(8)(1.06) + 2.68 = 11.16$ kips/inch of weld.

The designer has two choices in laying out a suitable concentrically loaded fillet weld to transmit a given connection load P:

1. Select a weld size, determine the weld value q_a in kips per lineal inch of weld, and then determine the total length l of weld required:

$$l = \frac{P}{q_a}$$

In using Table 6.1, $q_a = \bar{q}D$ for the metal-arc process welds and for submerged-arc welds of size $\frac{3}{8}$ in. or less. For submerged-arc weld sizes greater than $\frac{3}{8}$ in., use footnote[a] to Table 6.1.

2. Alternatively, if a particular weld length is suggested by the geometry of the connected parts, the required weld value is determined:

$$q = \frac{P}{l}$$

The required weld size (D) in sixteenths may then be determined by dividing the calculated force per inch (q) by $\bar{q}$ as listed in Table 6.1. This procedure must be modified for submerged arc welds of size greater than $\frac{3}{8}$ in. by first subtracting the tabulated bonus and then dividing the calculated remaining force by the *metal-arc electrode* listing for $\bar{q}$.

In designing a double fillet-welded tee joint [Fig. 6.5(a)] the allowable shear force per inch of two fillet welds may exceed the allowable shear force per inch of stem of tee, in which case the latter would control the design and impose an effective upper limit on the useful size of the fillet welds. Fillet-welded tee connections to a single plate should always be welded on both sides because of the weakness in bending and susceptibility to shipping damage of a single fillet weld.

For single or double angles under static tension load, the AISCS, Sec. 1.15.3, does not require that fillet welds in end connections be disposed so as to balance the forces about the neutral axis of the member. However, for members in compression, or subject to repeated stress variation, it is recommended that the fillet welds be so placed as to balance the forces about the neutral axis and eliminate eccentricity. Two common cases occur, as illustrated in Fig. 6.8: case 1, comprised of two longitudinal welds, and case 2, with a transverse weld added.

Case 1

Assume all fillet welds to be the same size. Then the equilibrium conditions of forces and moments about the center-of-gravity axis of the angle member are given, respectively, as follows:

$$l = l_1 + l_2 = \frac{P}{q_a} \tag{6.4}$$

$$c_1 l_1 q_a = c_2 l_2 q_a \tag{6.5}$$

Fig. 6.8 Balancing fillet welds for angle connections.

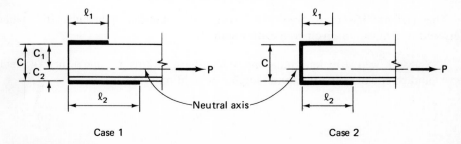

Case 1

Case 2

from which is obtained,

$$l_1 = \frac{c_2}{c} l \tag{6.6}$$

$$l_2 = \frac{c_1}{c} l \tag{6.7}$$

where l_1, l_2 = required lengths of fillet welds
$\quad\quad\quad l$ = required total lengths of fillet welds
$\quad\quad c_1$, c_2 = distances from neutral axis to extreme fibers of the angle
$\quad\quad\quad P$ = axial load
$\quad\quad\quad q_a$ = permissible strength of fillet weld

Case 2

If a weld of length c is provided along the end of the angle, the equilibrium conditions of forces and moments again lead to the following expressions:

$$l = l_1 + l_2 + c = \frac{P}{q_a} \tag{6.8}$$

$$c_1 l_1 q_a + c\left(\frac{c}{2} - c_2\right) q_a = c_2 l_2 q_a \tag{6.9}$$

Then

$$l_1 = \frac{c_2}{c} l - \frac{c}{2} \tag{6.10}$$

$$l_2 = \frac{c_1}{c} l - \frac{c}{2} \tag{6.11}$$

Concentrically loaded welded connection design will now be illustrated by means of Ex. 6.5, 6.6, and 6.7.

Example 6.5   Lap Joint (Welded)

Similar to Ex. 6.1, except plates are $\frac{1}{2} \times 8$ in. loaded to 50-kip tension. Determine the transverse weld size required, using A36 steel and E70 electrodes.

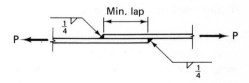

Solution

$$\text{minimum lap} = 5 \times \text{thickness of thinnest plate} \quad\quad \text{(AISCS 1.17.9)}$$
$$= 5 \times \tfrac{1}{2} = 2.5 \text{ in.}$$

total length of weld = approx. $8 \times 2 = 16$ in.

Approximate required weld force per inch:

$$q = \tfrac{50}{16} = 3.13 \text{ kips/in.}$$

From Table 6.1, $\frac{1}{16}$ E70 weld value $\bar{q}$ is 0.93 kips/in. Required weld size in $\frac{1}{16}s =$ 3.13/0.93 = 3.4; hence try $\frac{1}{4}$-in. fillet weld:

$$\text{weld value} = 4 \times 0.93 = 3.72 \text{ kips/in.}$$

Check minimum permitted weld size, AISCS, Table 1.17.5, which lists $\frac{3}{16}$ in., which is less than $\frac{1}{4}$ in. as used.

Example 6.6   *Butt Joint with Groove Weld*

A plate $\frac{1}{2} \times 12$ in. carrying a tensile force of 125 kips is to be spliced using a groove-welded butt splice, A36 steel, and E70 electrodes.

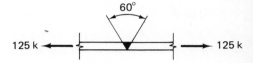

Solution

Actually, the only consideration here is to join the two pieces with a complete-penetration groove weld using electrodes that provide a weld with equal or greater tensile strength than that of the plate. A single vee complete-penetration groove weld is shown in sketch B-L2 as described on page 4-133 of the AISCM. Note that the weld root on the underside must be gouged and back welded. This is only one of several groove welds that could be used. For example, if the underside is inaccessible to the welder, B-U2 from page 4-134 of the manual could be used. This joint requires a backup strip to have been tack welded to one of the pieces prior to fit up.

Example 6.7

Same as Ex. 6.3 except that fasteners are replaced by welds. The bolted tee may now be replaced by a single $\frac{5}{8}$-in. plate† that is welded directly to the center of the column flange using a double-vee full-penetration groove weld, as shown in sketch B-U3a of the AISCM. Fillet welds attaching the angles to the plate should be arranged so as to balance the forces about the neutral axis of the connection of the double angles to eliminate any eccentricities.

A. Connection A. Assume that no weld is placed along the end of angles. Use maximum allowable weld size to connect angles to bracket. Then $w = \frac{1}{2} - \frac{1}{16} = \frac{7}{16}$-in. weld (AISCS, Sec. 1.17.6). The permissible strength of $\frac{7}{16}$-in. fillet weld (use E70 electrodes) is

$$q_a = 0.707wF_v$$
$$= 0.707 \times \tfrac{7}{16} \times 21 = 6.5 \text{ kips/in.}$$

---

†*Note:* As an alternative to the use of the $\frac{5}{8}$-in. plate, the WT 15 $\times$ 62 could again be used, as in Ex. 6.3, but the simplicity inherent to welded design would then be lost. It may be noted, however, that the protruding plate attached permanently in the shop to the column is a nuisance and subject to damage in shipment.

where F_v = permissible shear stress of E70XX (21 ksi), AISCS, Table 1.5.3.(Or use Table 6.1: $q_a = 7 \times 0.93 = 6.5$ kips/in.) Required total weld length $l = P/q_a = \frac{120/2}{6.5}$ = 9.23 in.

$$l_1 = \frac{c_2}{c} l = \frac{1.33}{4} 9.23 = 3.07 \text{ in. [Eq. (6.6)]}$$

$$l_2 = \frac{c_1}{c} l = \frac{2.67}{4} 9.23 = 6.17 \text{ in. [Eq. (6.7)]}$$

Use $l_1 = 3\frac{1}{4}$ in.; $l_2 = 6\frac{1}{2}$ in.

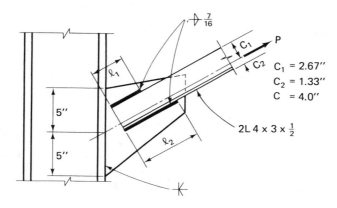

B. Connection B. According to AISCS, Table 1.5.3, the allowable stresses in tension and shear in a full penetration groove weld are the same as for the base metal. The AISCS does not deal explicitly with the problem of combined shear and tension in groove welds but an adequate design may be obtained by limiting the maximum principal tensile stress to the allowable stress in tension. As discussed later in Chapter 8, this may be accomplished by a closely approximate formula [Eq. (8.13)], which simply reduces the allowable tension stress in the direction of applied tensile force to a value F_{ra}:

$$F_{ra} = \left[1 - \left(\frac{f_v}{F_a} \right)^2 \right] F_a$$

The tension and shear components of the applied force in the groove weld, designated respectively as T and V, are:

$$T = P \cos 30° = 120 \times 0.866 = 103.9 \text{ kips}$$
$$V = P \sin 30° = 120 \times 0.5 = 60.0 \text{ kips}$$

Determine a trial groove weld length on the basis of an arbitrary reduction in F_a, say 18 ksi,

$$\text{Trial } l = \frac{103.9}{\frac{5}{8} \times 18} = 9.24 \text{ in.}$$

Try $l = 10$ in.

$$f_v = \frac{60}{\frac{5}{8} \times 10} = 9.6 \text{ ksi}$$

Then

$$F_{ra} = \left[1 - \left(\frac{9.6}{22} \right)^2 \right] 22 = 17.8 \text{ ksi}$$

$$f_a = \frac{103.9}{\frac{5}{8} \times 10} = 16.6 < 17.8 \qquad \text{OK}$$

($l = 9$ in. was tried and found insufficient).

6.5   ECCENTRICALLY LOADED CONNECTIONS

Previous design examples in this chapter have been limited to concentrically loaded connections. Concentricity is always desirable, but eccentrically loaded connections are sometimes required. Riveted, bolted, and welded connections will be considered.

When eccentricity of load imposes only shear force on riveted and bolted fasteners, with no change of initial tension, the assumption may be made that the eccentric load may be replaced by an equivalent force and couple acting at the centroid of the fastener group. If, on the other hand, both shear and tension are induced by load eccentricity, the design will require consideration of the following:

1. Reduced allowable tension stress in fasteners for bearing-type riveted or bolted connections, or

2. Reduced allowable shear stress in friction-type fasteners using high-strength bolts.

The foregoing effects were considered in Sec. 6.2.

Compact fillet-welded connections between heavy parts may be designed simply by providing a weld size adequate to resist the maximum resultant force per lineal inch of weld due to the combined effect of applied force and the moment induced by eccentricity.

Initial attention is now given to riveted or bolted shear connections that produce no tensile force change in the fasteners, as shown by the two-plate bracket bolted to the flanges of a W column in Fig. 6.9(a). The load P may be vertical, or inclined as shown, and there is an eccentric moment of magnitude Pa acting at the centroid of the fastener group, as shown in Fig. 6.9(b). In this illustration, because of the double symmetry of the fastener pattern, the centroid is readily located by inspection. It is convenient to replace P by its components P_x and P_y, as shown in Fig. 6.9(b). The problem is to determine the maximum resultant shear force on the particular most-stressed fastener, which usually can be located by elementary considerations.

In Fig. 6.9(c), R_{Ap} represents the resultant force due to applied load P acting on any *particular* bolt A. As assumed for concentric connections, $R_{Ap} = P/n$, where n is the total number of fasteners. R_{Am} is the resultant force due to moment, assumed to

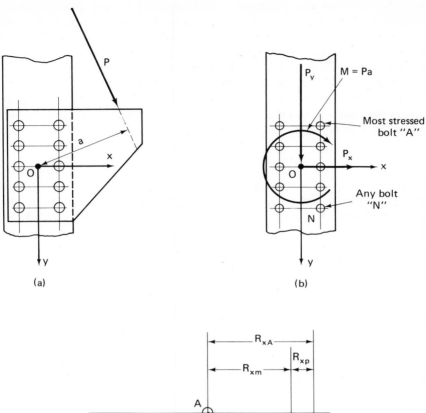

Fig. 6.9 Eccentrically loaded riveted or bolted connection.

be proportional to the radial distance r_A from centroid O, and acting normal to r_A. The total shear force on the fastener is shown as R_A, the resultant of R_{Ap} and R_{Am}, and is shown graphically as the diagonal of the force parallelogram. To avoid the determination of angles and the use of unwieldy arithmetic, it is convenient to break down each of these resultant forces into its x and y components.

We let R_{xp} and R_{yp} represent the shear force components in a single fastener due to P, respectively, in the x and y directions. Let R_{xm} and R_{ym} be the shear-force components in the same fastener due to M, respectively, in the x and y directions.

Consider P to be replaced by its components, P_x and P_y:

$$R_{xp} = \frac{P_x}{n} \quad \text{and} \quad R_{yp} = \frac{P_y}{n} \tag{6.12}$$

In Fig. 6.9(c) let N be *any* fastener at radial distance r_N from the centroid of the fastener group (O), and R_{Nm} the total force due to the eccentric moment. The contribution of fastener N to the moment resistance is equal to $R_{Nm}r_N$, and the total moment is equal to the sum of these individual contributions:

$$M = \sum R_{Nm}r_N \tag{6.13}$$

As assumed,

$$\frac{R_{Am}}{R_{Nm}} = \frac{r_A}{r_N} \quad \text{or} \quad R_{Nm} = \frac{R_{Am}r_N}{r_A}$$

Substituting this value of R_{Nm} into Eq. (6.13)

$$M = \frac{R_{Am}}{r_A} \sum r_N^2 \tag{6.14}$$

or, the force R_{Am}, due to moment, on any particular fastener is equal to

$$R_{Am} = \frac{Mr_A}{\sum r_N^2} \tag{6.15}$$

Breaking R_{Am} into its x and y components, and noting, by similar triangles, that

$$\frac{R_{xm}}{R_{Am}} = \frac{y_A}{r_A}, \qquad \frac{R_{ym}}{R_{Am}} = \frac{x_A}{r_A}$$

then

$$R_{xm} = \frac{y_A R_{Am}}{r_A} = \frac{My_A}{\sum r_N^2} \tag{6.16a}$$

$$R_{ym} = \frac{x_A R_{Am}}{r_A} = \frac{Mx_A}{\sum r_N^2} \tag{6.16b}$$

In calculating $\sum r_N^2$ it is convenient to use x and y component distances:

$$\sum r_N^2 = \sum x_N^2 + \sum y_N^2 \tag{6.17}$$

The x and y components of total shear force on fastener A due to both applied load and moment are

$$R_{xA} = R_{xp} + R_{xm} \tag{6.18a}$$

$$R_{yA} = R_{yp} + R_{ym} \tag{6.18b}$$

Finally, the resultant force on the most-stressed fastener is determined:

$$R_A = \sqrt{R_{xA}^2 + R_{yA}^2} \tag{6.19}$$

and the connection design is adequate if

$$R_A(\text{max}) \leq R(\text{allowed})$$

In Fig. 6.9 it is obvious by means of the following reasoning that the most-stressed bolt is at A:

1. R_{xp} and R_{yp} act in the positive x and y directions.

2. A is one of four locations where R_m is maximum.

3. A is the only bolt for which both R_{xm} and R_{ym} act in the same direction and thereby increase the magnitude of R_{xp} and R_{yp}.

Turning now from the riveted or bolted connection to the eccentrically loaded fillet-welded connection, the analysis is essentially similar and will not be developed in as much detail. In place of any single fastener, one now considers a differential length of fillet weld dl, and in place of $\sum r_N^2$ (which is the polar moment of inertia of the pattern of unit bolt areas) we substitute for a unit width of fillet weld throat

$$\int r_N^2 \, dl = I_z$$

The calculation of I_z can best be illustrated by means of an Ex. 6.9. Usually it is useful to take advantage of the relationship $I_z = I_x + I_y$.

Figure 6.10 shows an eccentrically loaded plate bracket welded to a column flange to provide a connection similar to the bolted bracket in Fig. 6.9. The same-sized fillet weld is assumed along the three edges of the bracket plate and one hidden edge of the column flange. Let the component x and y directed forces per unit length of weld at location A and due to applied load P be

$$q_{xp} = \frac{P_x}{l} \tag{6.20a}$$

$$q_{yp} = \frac{P_y}{l} \tag{6.20b}$$

l represents the total length of fillet weld measured at the root location.

Fig. 6.10 Eccentrically loaded welded connection.

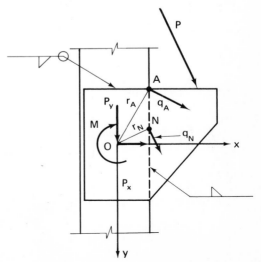

The component forces at the same location due to moment can be shown to be

$$q_{xm} = \frac{My_A}{I_z} \tag{6.21a}$$

$$q_{ym} = \frac{Mx_A}{I_z} \tag{6.21b}$$

I_z is the moment of inertia of the unit width weld about the z axis (through O in Fig. 6.10) and normal to the xy plane. As in the case of the bolted connection,

$$q_x = q_{xp} + q_{xm} \tag{6.22a}$$

$$q_y = q_{yp} + q_{ym} \tag{6.22b}$$

Thus the resultant force per unit length of weld—the determining factor in selecting the weld size—is

$$q_r = \sqrt{q_x^2 + q_y^2} \tag{6.23}$$

The foregoing procedure may be readily extended to a general three-dimensional fillet-welded joint with the addition of a q_z force component.

Riveted and bolted connections in which eccentricity induces change in fastener tension will be considered in Sec. 6.7 under the heading of moment-resisting beam-to-column-type connections. Examples 6.8 and 6.9 illustrate the design of shear-type connections using high-strength bolts and welds, respectively.

Example 6.8 *Eccentric Loads on High-Strength Bolted Bracket*

Design a bracket connected to the faces of a W 14 × 184 column to support a girder reaction of 120 kips applied eccentrically, as shown. Using four lines of bolts

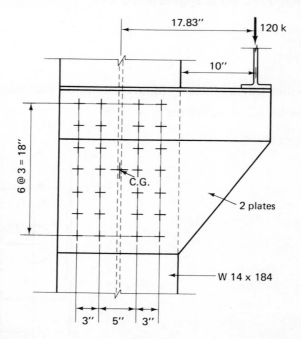

in each face of the column, determine the number and size of A490 high-strength bolts in friction-type connection in accordance with the AISCS.

Solution

Assume 7 bolts with 3-in. spacing in each vertical line. The centroid of the bolt group is at point O from symmetry. Shear stresses in bolts shall be in accordance with AISCS, Table 1.5.2.1.

$$P_x = 0$$

$$P_y = \frac{120}{2} = 60 \text{ kips}$$

$$M_z = 120 \times \frac{17.83}{2} = 1070 \text{ kip-in.}$$

$$n = 7 \times 4 = 28 \text{ bolts}$$

$$\sum r_N^2 = \sum (x_N^2 + y_N^2) = 14(5.5^2 + 2.5^2) + 8(9^2 + 6^2 + 3^2) = 1519 \text{ in.}^2$$

By inspection, the bolt in the right-hand upper corner is the most-stressed bolt. Compute the resultant shear force R on the upper-right-corner bolt where $x = 5.5$ in. and $y = 9$ in.; then according to Eqs. (6.12), (6.16), (6.17), and (6.19),

$$R_x = \frac{P_x}{n} + \frac{M_z y}{\sum (x_N^2 + y_N^2)} = 0 + \frac{1070 \times 9}{1519} = 6.34 \text{ kips/bolt} \longrightarrow$$

$$R_y = \frac{P_y}{n} + \frac{M_z x}{\sum (x_N^2 + y_N^2)} = \frac{60}{28} + \frac{1070 \times 5.5}{1519} = 6.01 \text{ kips/bolt} \downarrow$$

$$R = \sqrt{R_x^2 + R_y^2} = \sqrt{6.34^2 + 6.01^2} = 8.74 \text{ kips/bolt}$$

Try $\frac{3}{4}$-in. bolts; the shear allowable stress in single shear is

$$A_b F_v = 0.44 \times 20 = 8.84 \text{ kips/bolt} > R_{\max} \qquad \text{OK}$$

where A_b = nominal body area of a bolt
            F_v = allowable shear stress on bolts, see AISCS, Table 1.5.2.1

(Alternatively, 8.84 is listed on page 4-4 of AISCM.)
Use 28 A490 high-strength bolts of $\frac{3}{4}$-in. diameter. Note that the resultant stress on the lower right bolt is of the same magnitude, but the direction is changed, since R_x acts toward the left. ←

Reference should be made to the AISCM, pages 4-60 and 4-65, which provide design tables for this type of connection using a reduced effective distance in calculating the eccentric moment. In other respects, the procedure is the same as the foregoing, and is justified on the basis of comparison with test results.

Example 6.9

For a single plate bracket welded to the face of a column, as shown in the figure, determine the size of the fillet welds required. Use E70 electrodes and A36 steel.

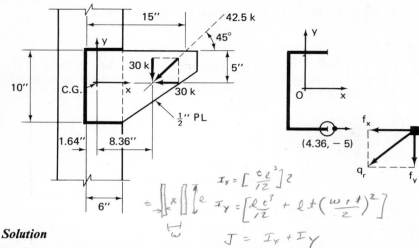

$$I_x = \left[\frac{t\ell^3}{12}\right]2$$

$$I_y = \left[\frac{\ell t^3}{12} + \ell t\left(\frac{w+t}{2}\right)^2\right]$$

$$J = I_x + I_y$$

Solution

Find the center of gravity of the weld root line. Vertically, by inspection, it is 5 in. from the top or bottom of the plate, because of symmetry. Horizontally, $(2 \times 6 \times 3)/22 = 1.64$ in.

Design loads acting at center of weld group:

$$P_x = 42.5 \cos 45° = 30 \text{ kips}$$

$$P_y = 42.5 \sin 45° = 30 \text{ kips}$$

$$P_z = 0$$

$$M_x = M_y = 0$$

$$M_z = 30 \times 8.36 = 251 \text{ kip-in.}$$

Moment of inertia:

$$I_x = 2 \times 6 \times 5^2 + \tfrac{1}{12} \times 10^3 = 384 \text{ in.}^4$$

$$I_y = 10 \times 1.64^2 + 2[\tfrac{1}{3} \times (1.64^3 + 4.36^3)] = 85 \text{ in.}^4$$

$$I_z = I_x + I_y = 384 + 85 = 469 \text{ in.}^4$$

Note the shearing stresses on the sketch, which indicate that the lower right edge (where $x = 4.36$, $y = -5$ and all stress components are additive) is the most highly stressed:

$$q_x = \frac{P_x}{l} + \frac{M_z y}{I_z} = \frac{30}{22} + \frac{251 \times 5}{469} = 1.36 + 2.68 = 4.04 \text{ kips/in.} \leftarrow$$

$$q_y = \frac{P_y}{l} + \frac{M_z x}{I_z} = \frac{30}{22} + \frac{251 \times 4.36}{469} = 1.36 + 2.33 = 3.69 \text{ kips/in.} \downarrow$$

may shear

$$\boxed{q_r = \sqrt{q_x^2 + q_y^2}}$$

$$= \sqrt{4.04^2 + 3.69^2} = 5.48 \text{ kips/in.}$$

From Table 6.1, $\frac{1}{16}$ E70 weld value $\bar{q} = 0.93$ kip/in. Weld size required in $\frac{1}{16}s = 5.48/0.93 = 5.9$. Use $\frac{3}{8}$-in. weld.

6.6 SHEAR CONNECTIONS FOR BUILDING FRAMES

A variety of beam-to-column or beam-to-girder connections is available to support simple beam reactions. They are purposely made flexible with regard to rotation between the ends of the beam and the column or girder. These are designated by AISCS as Type 2 connections and are used in structures for which lateral forces do not need to be considered, or where other bents in the building resist wind and seismic forces by frame action, truss framing, or shear walls. Flexible connections for reactions may involve attachment only to the beam web, as shown in Fig. 6.11, or may consist of top and bottom angles, respectively designated by AISCS as "framed beam connections" and "seated beam connections." They may include rivets, bolts, or welds, alone or in combination. The AISCM, pages 4-12 to 4-59, provides descriptive and tabular design information covering the most commonly used types. These connections do develop a certain amount of moment, which may amount to 10 per cent of the full fixed end moment or even more. However, these moments are disregarded in design.

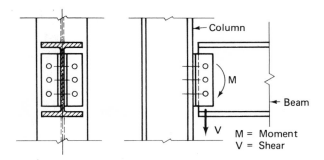

Fig. 6.11   Framing angle connection.

In the case of the riveted or bolted stiffened seat angle, the fasteners that attach the angles to the column or girder are in combined shear and tension due to the eccentricity of the applied load. The bending moment is transmitted through the connection by tension in the upper fasteners and by bearing pressure in the lower part between the angles and the column, as shown by the shaded area in Fig. 6.12. The equivalent effective area in transmitting the bending moment is shown in Fig. 6.12(b) in which area ac_1 equals the area of the eight fasteners and whose stress distribution, due to moment, is shown in Fig. 6.12(c). Then

$$a = \frac{mA}{p} \tag{6.24}$$

$$c_1 = \frac{\sqrt{b}}{\sqrt{a} + \sqrt{b}} h \tag{6.25}$$

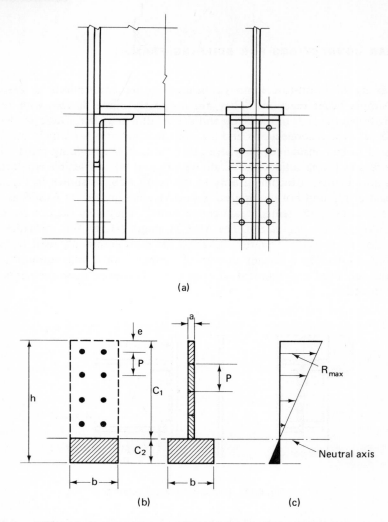

Fig. 6.12 Seat angle connection with fasteners subjected to tension: (a) actual stress portions; (b) equivalent cross-section; (c) stress distribution.

$$R_{\max} = \frac{M}{I}(c_1 - e)A \qquad (6.26)$$

where a = width of the equivalent fastener area
p = pitch of the fasteners
m = number of the fasteners per horizontal row
A = cross-sectional area of a fastener
c_1, c_2 = distance between neutral axis and extreme fibers
b = width of the framing angles or tee

h = height of the framing angles or tee

R_{max} = maximum tensile load in a fastener

I = moment of inertia = $(ac_1^3 + bc_2^3)/3$

e = edge distance between the maximum tensile load of the fastener and the edge of the framing angles

The condition shown in Fig. 6.12 applies only to the specialized case in which we have a uniform vertical spacing of the fasteners. However, this happens very frequently in practice, and the equations that were developed are very simple to use.

If the vertical spacing of the fasteners varies, we would have a varying thickness (a) that could not be used in these equations. Therefore, we must use another method in which we consider the moments of the individual fasteners about the neutral axis. The neutral axis usually lies between one sixth to one seventh of the length of the connection (h) from the bottom of the connection. This assumption is only used to estimate the number of fasteners in tension (above the neutral axis). An equation is then written in which the moment of the compression area about the neutral axis is equated to the moment of the areas of the tension fasteners about the neutral axis, since these must be equal. The resulting answer will indicate whether we have made a correct assumption.

Once the neutral axis is located we can calculate the moment of inertia. Next, using the flexure formula, we can determine the tensile stress in the critical fasteners and the compression stress on the extreme fiber of the connection.

An application of the preceding procedure is illustrated subsequently in Ex. 6.12. Other connection design principles required in the examples of beam shear connections that follow are based on procedures that have been covered previously in this chapter, and maximum use will be made of the design-aid tables in the AISCM. Tables I and II, pages 4-12 to 4-26 of the manual, are to be used for riveted and bolted connections. Table III, starting on page 4-28, is for use when angles are welded to the beam web, in conjunction with Table I for determining the number of rivets or bolts to be used in the outstanding legs. Table IV is to be used for connections welded both in the shop and field, using E70 electrodes. Tests have shown that these standard framed connections do have high rotation capacity and will support the tabulatd loads.

The examples that follow are supplementary to those in Part 4 of the AISCM, which should be studied carefully.

Example 6.10   *Framed Beam Shear Connections*

A. Bearing-type connection

An end connection for a W 18 × 64 beam is subjected to a reaction of 50 kips. Material for beam and connections is ASTM A36. Use $\frac{3}{4}$-in.-diameter A325 bolts in bearing-type connection, with threads not excluded from shear plane.

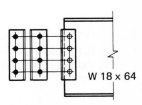

W 18 × 64

Solution

thickness of beam web = 0.403 in.

Refer to Table I-A4, AISCM, page 4-20.

capacity of connection = 53 kips > 50   OK

angle thickness = $\frac{1}{4}$ in.

Refer to AISCM, Table I-B4.

total bearing capacity if web is 1 in. thick = 146 kips

bearing capacity of fasteners on beam web = 146 × 0.403 = 59 kips > 50   OK

Use two angles 4 × $3\frac{1}{2}$ × $\frac{1}{4}$ × $11\frac{1}{2}$ in. long with four bolts in beam web and eight bolts in outstanding legs.

Note: Referring to AISCM, Table I-A4, it is seen that the same number of fasteners and angle size would be required if A325 bolts were used in friction-type connections or if A502-1 rivets were used.

B. Friction-type bolted connection

An end connection for a W 24 × 130 beam is subjected to a reaction of 175 kips. Material for beam and connection is ASTM A36. Use $\frac{7}{8}$-in.-diameter A325 bolts in friction-type connection in beam web and in the outstanding legs.

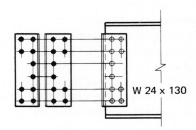

Solution

The longest angle that can be used in a 24-in. beam web is 1 ft-$8\frac{1}{2}$ in., and therefore we can only use a maximum of seven fasteners per line. With the fasteners specified, the maximum capacity of the connection listed in AISCM, Table I-A7 (page 4-18), is 126 kips. We shall therefore be required to use two lines of bolts per leg, and using AISCM, Table II-A6, page 4-24,

capacity of connection = 180 kips > 175   OK

angle thickness = $\frac{3}{8}$ in.

Bearing stress is not restricted and need not be checked in friction-type high-strength bolted connections (AISCS, Sec. 1.5.2.2). Use two L6 × 6 × $\frac{3}{8}$ × 1 ft $5\frac{1}{2}$ in. long with 10 fasteners in each leg.

C. Shop welded field bolted connection

Same as (B) except weld angles to beam web and use A325 $\frac{7}{8}$ in. bolts in bearing-type connection with threads excluded from shear plane on outstanding legs.

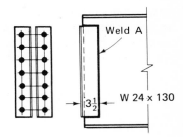

Weld A

$3\frac{1}{2}$

W 24 x 130

Solution

Refer to AISCM, Table I-A7, page 4-18:

$$\text{capacity of connection} = 185 \text{ kips}$$

$$\text{angle thickness} = \tfrac{5}{16} \text{ in.}$$

Refer to AISCM, Table III, weld A for angle 1 ft 8 $\frac{1}{2}$ in. long with $\frac{5}{16}$ in. weld; the capacity is 186 kips for a minimum beam web thickness of 0.53 in. (for a W 24 $\times$ 130, $t_w = 0.565$ in. 0.53, hence weld capacity A need not be reduced). Minimum angle thickness $= \frac{5}{16} + \frac{1}{16} = \frac{3}{8}$ in. (See footnote [a], AISCM page 4-31.) Bearing stress in outstanding legs of angles must be checked. The capacity of seven bolts in a 1-in. plate is 298 kips (AISCM, Table I-B7, page 4-18). Hence 14 bolts in a $\frac{3}{8}$-in. plate have a bearing capacity of

$$298 \times 2 \times \tfrac{3}{8} = 223 \text{ kips} > 175 \qquad \text{OK}$$

Use two L 4 $\times$ 3$\frac{1}{2}$ $\times$ $\frac{3}{8}$ $\times$ 1 ft 8$\frac{1}{2}$ in. long with seven fasteners per row in outstanding legs and $\frac{5}{16}$-in. weld connecting angles to beam web.

D. All welded connection

Same as (B) except use E70 weld for connection of both web and outstanding legs of angles.

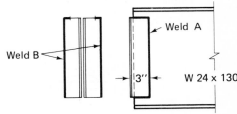

Weld A

Weld B

3"

W 24 x 130

Solution

Refer to Table IV, AISCM, page 4-36.

weld A: use $l = 20$ in., $\frac{5}{16}$ in. weld good for 182 kips

weld B: use $l = 20$ in., $\frac{3}{8}$ in. weld good for 181 kips

minimum web thickness $= 0.53$ in. < 0.565

Use two L 4 $\times$ 3 $\times$ $\frac{7}{16}$ $\times$ 1 ft-8 in. long.

E. Welded end plate connection—field bolted

This type of connection, consisting of a plate welded to the beam web with fillet welds on both sides, compares favorably to the double-angle connection in providing rotation capacity with end plate thicknesses of $\frac{1}{4}$ to $\frac{3}{8}$ in. The distance between lines of fasteners should be from $3\frac{1}{2}$ to $5\frac{1}{2}$ in., and the edge distance should not be greater than $1\frac{1}{4}$ in. to minimize prying action on the plate and fasteners. This example is the same as (A) except use E70 welded end plate shear connection with $\frac{3}{4}$-in.-diameter A325 bolts in friction-type connection.

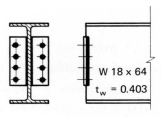

W 18 x 64

$t_w = 0.403$

Solution

Refer to Table IX, AISCM, page 4-59.

$$\text{capacity of 8 fasteners} = 53 \text{ kips} > 50 \qquad \text{OK}$$

$$\text{minimum plate thickness} = 0.182 \text{ in.}$$

$$\text{length of plate} = 11\frac{1}{2} \text{ in.}$$

for a minimum web thickness of 0.389 in. with $F_y = 36$ ksi and $\frac{3}{16}$-in. weld,

$$\text{weld capacity} = 61.9 \text{ kips} > 50 \qquad \text{OK}$$

Use $\frac{1}{4} \times 8 \times 11\frac{1}{2}$ in. plate connected to beam web with two lines of $\frac{3}{16}$-in. fillet weld and eight $\frac{3}{4}$-in. A325 bolts for the field connection.

Example 6.11 *Seated Beam Connection (Riveted, Bolted, or Welded)*

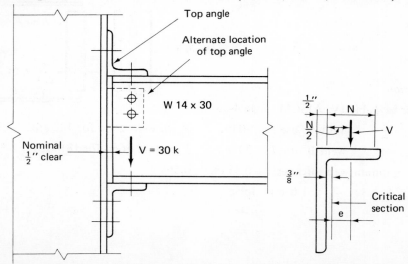

Solution

For W 14×30, $b_f = 6.73$ in., $t_w = 0.27$ in., and $k = 1.0$ in.
Length of bearing required:

$$N + k = \frac{R}{0.75F_yt_w} \qquad \text{(AISCS Formula 1.10-9)}$$

then
$$N = \frac{30}{0.75 \times 36 \times 0.27} - 1.0 = 3.1 \text{ in.}$$

Use 4-in. leg, 8 in. long. Assume $\frac{5}{8}$-in. angle thickness (governed by bending); the distance from back of angle to the critical section in bending is $\frac{5}{8} + \frac{3}{8} = 1$ in. It is usually considered that the load is concentrated at the center of required bearing (3.1 in. in this example):

$$\text{eccentricity of load} = 0.5 + \tfrac{3.1}{2} - 1.0 = 1.05 \text{ in.}$$
$$M = 30 \times 1.05 = 31.5 \text{ kip-in.}$$
$$F_b = \frac{6M}{bt^2} = \frac{6 \times 31.5}{8t^2}$$

then
$$t^2 = 0.875 \qquad (F_b = 27 \text{ ksi}) \quad \text{and} \quad t = 0.935 \text{ in.}$$

Try $\frac{7}{8}$-in. thickness (eccentricity $= 0.80$ in. and $M = 24$ kip-in.), and by calculations paralleling the preceding, the required

$$t = 0.82 \text{ in.} < 0.875 \qquad \text{OK}$$

Use L $6 \times 4 \times \frac{7}{8} \times 8$ in. long.

Alternative procedures for attachment to the column are:

A. Riveted: $\frac{7}{8}$-in. A502 Grade 1 rivets good for 9.02 kips/rivet (from AISCM, page 4-6):

$$n = \frac{30}{9.02} = 3.33 \qquad \text{use 4 rivets}$$

B. Bolted: $\frac{3}{4}$-in. A325 bolts in bearing-type connection with threads excluded from shear plane, good for 9.72 kips/bolt (from AISCM, page 4-4):

$$n = \frac{30}{9.72} = 3.09, \qquad \text{use 4 bolts}$$

Connections of the beam to the seat angle, beam to top angle, and top angle to column can be made using two $\frac{3}{4}$-in.-diameter unfinished bolts. The top angle simply holds the beam in a vertical position and must be flexible to permit simple beam end rotation. Use L $4 \times 4 \times \frac{1}{4} \times 8$ in. long.

C. To use a weld to attach seat angles to column, assume 6-in. vertical welds on the edge of the seat angle as shown.

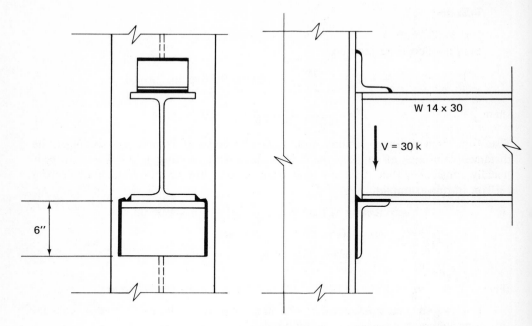

$$\text{moment on weld} = \frac{30}{2}\left(\frac{1}{2} + \frac{3.1}{2}\right) = 30.8 \text{ kip-in.}$$

$$q_h = \frac{Mc}{I} = \frac{30.8 \times 3.0}{18.0} = 5.13 \text{ kips/in.} \quad \text{(using unit thickness of weld)}$$

$$q_v = \frac{30}{2 \times 6} = 2.5 \text{ kips/in.}$$

$$q_r = \sqrt{5.13^2 + 2.5^2} = 5.72 \text{ kips/in.}$$

$$\text{size of weld reqd} = \frac{5.72}{0.93} = 6.2\left(\frac{1}{16}\right)s$$

Use $\frac{7}{16}$ in. weld. (See also the alternate procedure in AISCM, page 4-45, Table VI-C.) For top angle use L $4 \times 4 \times \frac{1}{4}$ in. angle, 6 in. long. Use $\frac{1}{4}$-in. welds on toes of angles as shown. Use $\frac{5}{16}$ in. welds 2 in. long on each side of lower flange of beam for attachment to seat angle.

Example 6.12

Design a riveted stiffened seat connection for a W 18×45 for an end reaction of 40 kips. All material A36 steel. Refer to the figure. Sketches (b) and (c) give additional details of the seat, and sketch (d) is for use in calculating rivet stress.

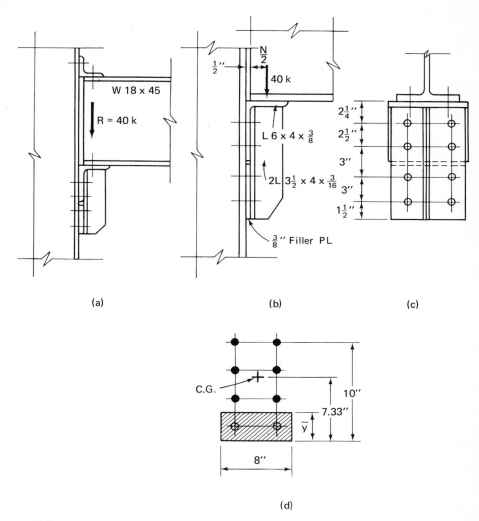

(a)   (b)   (c)

(d)

Solution

For W 18×45, $b_f = 7.5$ in., $t_w = 0.335$ in., and $k = 1.0$ in.:

$$N = \frac{40}{27 \times 0.335} - 1.0 = 3.42 \text{ in.}$$

Use 4-in. leg, which allows a 0.59-in. setback. Eccentricities of load for stiffened seats will be assumed to be a nominal setback of 0.5 in. plus half the remaining seat length.

Assumed eccentricity of end reaction:

$$e = 0.5 + \frac{3.5}{2} = 2.25 \text{ in.}$$

$$\text{moment} = 2.25 \times 40 = 90 \text{ kip-in.}$$

Try $\frac{3}{4}$-in. A502 Grade 1 rivets.

Assume neutral axis to be above the two lowest rivets and locate center of gravity (c.g.) of upper 6 rivets. From the bottom of vertical angles to c.g. is

$$4.5 + \frac{(2 \times 3) + (2 \times 5.5)}{6} = 7.33 \text{ in.}$$

Refer to sketch (d):

$$8\bar{y}\,\frac{\bar{y}}{2} = 6 \times 0.442(7.33 - \bar{y})$$

$$4\bar{y}^2 + 2.65\bar{y} - 19.4 = 0$$

Solving, $\bar{y} = 1.90 \text{ in.}$

$$I = 8 \times \frac{1.90^3}{3} + 2 \times 0.442(2.6^2 + 5.6^2 + 8.1^2) = 110 \text{ in.}^4$$

Tensile stress in the top rivets is $90 \times 8.1/110 = 6.63$ ksi.

$$\text{shear stress } f_v = \frac{40}{8 \times 0.442} = 11.3 \text{ ksi}$$

Allowable tensile stress (AISCS, Sec. 1.6.3) is

$$F_t = 28.0 - 1.6f_v = 28.0 - 1.6 \times 11.3 = 9.9 \text{ ksi} > 6.63 \quad \text{OK}$$

The required bearing area of stiffeners is $\frac{40}{27} = 1.48$ in.^{2}, and the area furnished (using L $4 \times 3\frac{1}{2} \times \frac{5}{16}$ in. stiffeners) is $(3.5 - 0.5)\frac{5}{16} \times 2 = 1.87$ in.^{2}.

For an alternate design procedure making use of tabular aids, refer to AISCM, pages 4-47 through 4-49.

Example 6.13

Redesign the stiffened seat of Ex. 6.12 as all welded, using A36 steel and E70 electrodes. Use a structural tee section, 4 in. in length, as shown.

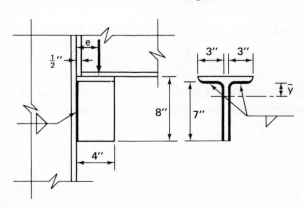

End eccentricity and moment are the same as for Ex. 6.12. Try WT 8 × 29 (has a wider flange than beam, to permit downhand fillet welds between beam and seat). Assume effective weld lengths as shown in the sketch, and locate c.g. of weld pattern (assume unit width of weld):

$$\bar{y} = \frac{3.5 \times 14.0}{20} = 2.45 \text{ in.}$$

$$I_x = 6 \times 2.45^2 + \tfrac{2}{3}(2.45^3 + 4.55^3) = 108.6 \text{ in.}^4$$

Assume bearing of stem on column at bottom, making the resultant weld stress at top the critical value. Due to eccentric moment,

$$q_H = \frac{90.0 \times 2.45}{108.6} = 2.0 \text{ kips/in.}$$

$$q_V = \tfrac{40}{20} = 2.0 \text{ kips/in.}$$

$$\text{resultant } q_R = \sqrt{2.0^2 + 2.0^2} = 2.83 \text{ kips/in.}$$

$$\text{weld size in } \tfrac{1}{16} s = \frac{2.83}{0.93} = 3.0 \qquad \text{(Table 6.1)}$$

Although $\tfrac{3}{16}$-in. welds are adequate for strength, $\tfrac{1}{4}$-in. must be used to meet the requirements of AISCS Table 1.17.5, page 5-44. Still greater size will be required if the column flange thickness exceeds $\tfrac{3}{4}$ in.

6.7 MOMENT-RESISTING CONNECTIONS

The two types of moment-resisting connections classified in AISCS, Sec. 1.2, as "rigid" (Type 1) and "semi-rigid" (Type 3) were discussed in Sec. 6.1.

Rigid connections are used in continuous-frame construction that resists lateral wind and seismic forces in tall buildings and are used occasionally in industrial building frames. Rigid connections are also a requirement in the construction of frames that are proportioned according to plastic design, in which case the connections must be strong enough to develop complete moment yielding at adjacent *plastic hinges*. Rigid connections are always advantageous if a building is loaded accidentally by explosive blast, earthquake, or high wind beyond its intended normal-use load. In such cases continuous-frame behavior, whether or not proportioned by plastic design, will provide a residual strength against ultimate collapse that may be a life-saving feature.

Semirigid connections are used in semi-continuous-frame construction, primarily for gravity load resistance, and primarily in office or apartment buildings of moderate height. The concept is intended to provide an economical balance between simple beam design, for which the maximum bending moment at the center is $0.125 \ wL^2$, and fully continuous construction, for which the maximum moment in the fully fixed end condition is $0.083 \ wL^2$ at the ends of the beam. In a fixed-ended beam with semirigid connections, the end moments and center moments for uniform gravity load could, ideally, be balanced at $0.0625 \ wL^2$, thus achieving savings in weight of beam and cost of connections that are designed for less than full continuity. However, the careful balance that is required between strength and flexibility is difficult to standardize and control in semirigid connections; furthermore, the advent of plastic design and the development of fully continuous welded construction have lessened the economic advantage of the semicontinuous construction. In current practice it is customary to label a connection as "semirigid" if it is designed for some specific moment capacity at any level less than that required to develop full continuity.

Two detailed examples of rigid-connection design are provided in the AISCM. In the first, on page 4-92, the moment is resisted by flange plates that are shop welded to the column and fastened to the beam flanges by high-strength bolts in the field. The shear is transferred from beam web to column by a vertical plate shop welded to the column and field fastened to the beam. The second example, starting on page 4-96, utilizes an end plate fillet welded to the beam and field bolted to the column. In end plate construction, careful dimensional shop control of the overall distance between the end plate faces is required to maintain proper column alignment. Alternatively, shims may be supplied, and an underrun tolerance allowed. (See Fig. 3.4 for illustration of end plates.)

An example of semi-rigid-connection design is provided on pages 4-88 to 4-91 of the AISCM. Refer to the illustration on page 4-88. Moment resistance is provided by the top and bottom plates that are fillet welded to the beam flanges and joined to the column flanges with single vee groove welds provided with backup bars so as to qualify the welds as full penetration. Adjacent to the column, a length of top plate equal to 1.5 times the plate width is provided to allow elongation under load so as to prevent the connection from being too rigid without the strength necessary for full moment continuity. Shear at the end is transferred from beam web to column by a vertical plate shop welded to the column and field bolted to the beam. This plate also serves to support the beam in position while the welds are being made during construction. By increasing connecting plate sizes and welds, this Type 3 connection could be modified so as to qualify as a rigid Type 1 connection.

Figure 6.13 shows a riveted or bolted moment and shear connection that also may be designed either as semirigid or rigid. It is a type that was much used prior to

Fig. 6.13 Moment and shear connection.

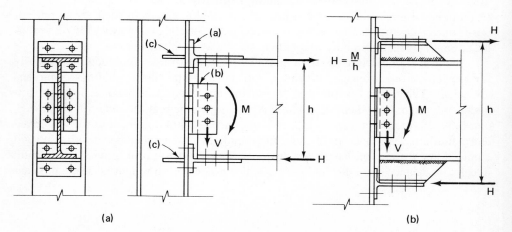

(a)

(b)

the advent of welding, and here serves to supplement the three examples provided by the AISCM. In addition, it introduces the problem of prying action added to bolt or rivet force. Another topic that will be covered briefly concerns the required strengthening of the column web against local deformation adjacent to the beam flanges. Although moment-resisting welded semirigid and rigid connections are by far the most commonly used in current practice, design examples are omitted herein simply because of the very adequate coverage in the AISCM.

In the connection shown in Fig. 6.13(a) the bending moment (M) is transmitted to the column flange by two tees (a) on the top and bottom flanges of the beam by tension and compression loads (H), respectively. The shear load (V) is transmitted to the columns by the two angles (b) connected to the beam web. Then

$$H = \frac{M}{h}$$

where h is the depth of the beam. It should be noted that the moment capacity of this connection can be increased by increasing the distance between the tees, as shown in Fig. 6.13(b). This greater distance will decrease the magnitude of force (H).

The top flange moment connection shown in Fig. 6.13(a) transmits the applied moment to the column by means of fasteners acting in tension. Consequently, the connecting element (the tee section) is subjected to a bending stress and deforms, as shown in Fig. 6.14, thereby creating a prying action. This prying action in turn creates a pressure (prying force ΔH) at the outer edge of the tee and also produces an additional external tension on the fasteners. Refer also to the three-page discussion in the AISCM, pages 4-80 to 4-82, which illustrates application of the same concept and use of the following formulas in the design of hanger-type connections.

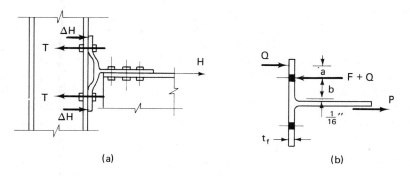

(a)   (b)

Fig. 6.14   Prying action in connection.

For connections using A325 bolts,

$$Q = \frac{100bd_b^2 - 18wt_f^2}{70ad_b^2 + 21wt_f^2} F \qquad (6.27)$$

For connections using A490 bolts,

$$Q = \frac{100bd_b^2 - 14wt_f^2}{62ad_b^2 + 21wt_f^2} F \qquad (6.28)$$

where $\quad P =$ allowable load on two angles or a structural tee, in kips per lineal inch, as determined by stress due to bending, for allowable $F_b = 0.75F_y$

$Q =$ prying force per fastener, in kips

$F =$ externally applied load per fastener $(wP/2)$, in kips

$w =$ length of flange tributary to each bolt, in inches

$d_b =$ nominal bolt diameter, in inches

$a =$ distance from fastener line to edge of flange, but not to exceed $2t_f$, in inches

$t_f =$ flange thickness, in inches

$b =$ distance from face of web to bolt line minus $\frac{1}{16}$ in. in inches

By referring to the formulas it can be seen that when the flanges are thick and the gage lines are closely spaced, the prying action will be very small. In the preliminary selection of flange thickness it may be assumed that a point of inflection exists at the midlength of dimension b, in which case the flange moment per lineal inch is

$$M = \frac{Pb}{4} \tag{6.29}$$

In the final check of flange bending adequacy, the maximum moment may be assumed to be the greater of two values: At the fastener line, the total moment per fastener

$$M_2 = Qa \tag{6.30}$$

and $\frac{1}{16}$ in. from the face of the tee stem,

$$M_1 = (F + Q)b - Q(a + b) \tag{6.31}$$

The foregoing nomenclature corresponds to that used in the AISCM.

The design of column web stiffeners, [plates (c) in Fig. 6.13] is covered by AISCS, Sec. 1.15.5, page 5-40. The use of the applicable formulas [AISCS, (1.15-1) through (1.15-4)] is illustrated by AISCM information on moment connections, pages 4-88 to 4-99. The requirements for column web stiffeners are based on the assumption that the beam is fully restrained, in which case the connection will develop the full yield point of the beam flange without failure and that the force H at yield will be balanced by yielding in the column web over a spread distance of $(t_b + 5k)$, as expressed by the following relationship:

$$F_{y(\text{col.})}t(t_b + 5k) \geq F_{y(\text{beam})} A_f \geq H$$

Solving the foregoing for required t for no stiffeners leads to AISCS Formula 1.15-1, page 5-40. As pointed out in the AISCS Commentary, this procedure is overconservative when the full restraint moment is not required, as in a semi-rigid connection, in which case the requirement could be based on the required H instead of the H to produce flange yield, as follows;

$$t = \frac{H}{F_{yc}(t_b + 5k)} \tag{6.32}$$

where F_{yc} is the yield point of the column steel. If H is introduced by a tee, as in Fig.

6.13, the spread of force into the column web will be greater than if introduced by a direct flange or flange plate connection, thus tending to make Eq. (6.32) more conservative.

It should be noted in the following Ex. 6.14 that for design as a rigid connection the design moment should be approximately equal to the moment capacity of the beam. With a section modulus of 89.1, at a stress f_b of 24 ksi, the moment for full connection rigidity would be $89.1 \times \frac{24}{12} = 178.2$ kip-ft.

Example 6.14   *Semirigid High-Strength Bolted Beam to Column Connection*

The connection is to be designed for a shear of 54 kips and a bending moment of 100 kip-ft, similar to the connection shown in Fig. 6.13. See also the accompanying sketch. Use A325 high-strength bolts and A36 steel.

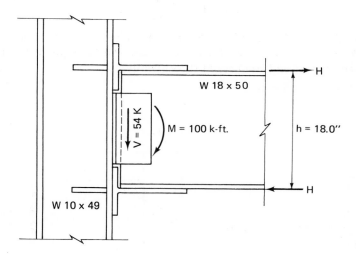

Solution

A. Moment connectors (use $\frac{7}{8}$-in. bolts in bearing-type connection)

Use two tee shapes at top and bottom flanges of the beam W 18×50 to carry the end moment. Then the tension and compression load H at top and bottom flanges of the beam are

$$H = \frac{M}{h} = \frac{100 \times 12}{18} = 66.7 \text{ kips}$$

(1) Flange tees to beam:

Allowable shear stress in single shear with threads excluded from shear planes:

$$F_v A_b = 22 \times 0.6013 = 13.23 \text{ kips/bolt}$$

$$\text{no. bolts reqd} = \frac{66.7}{13.23} = 5.04$$

Use six $\frac{7}{8}$-in. A325 high-strength bolts, and exclude threads from shear planes. Determine minimum web thickness of tee for allowable bearing stress of $1.35 \times 36 = 48.6$ ksi (AISCS, Sec. 1.5.2.2).

$$\text{force per bolt} = \frac{66.7}{6} = 11.1 \text{ kips}$$

$$11.1 = \tfrac{7}{8}(48.6)t \quad \text{whence} \quad t_{\min} = 0.26 \text{ in.}$$

(2) Tees to column flange:
Allowable tensile strength:

$$F_t A_b = 40 \times 0.601 = 24.05 \text{ kips/bolt}$$

where F_t = allowable tensile stress (AISCS, Table 1.5.2.1).

$$\text{no. reqd} = \frac{66.7}{24.05} = 2.8$$

Use four A325 high-strength bolts of $\frac{7}{8}$-in. diameter to join tees to column flange. *Note:* Extra bolt strength, as provided, will be required to take care of prying action. This will be checked subsequently.

(3) Required thickness for the tees:

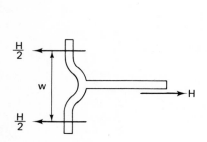 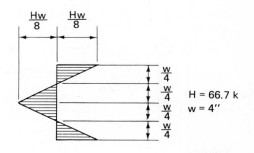

Try 8-in.-long WT (use 4-in. bolt gage = w).
Maximum moment in tee flange:

$$m = \frac{Hw}{8} = 66.7 \times \frac{4}{8} = 33.3 \text{ kip-in.}$$

Required thickness:

$$t = \sqrt{\frac{6m}{bF_b}}$$

where b = length of tee
F_b = allowable bending stress

$$t = \sqrt{\frac{6 \times 33.3}{8 \times 27}} = 0.962 \text{ in.}$$

Use 8-in.-long WT 10.5 $\times$ 63.5.

$$t_f = 0.985 > 0.962 \quad \text{OK}$$

$$t_w = 0.588 > 0.26 \quad \text{OK}$$

Check the tensile capacity of the net area of the tee web to transmit the force H of 66.7 kips:

$$H = 0.588 \times 22[8 - 2(\tfrac{7}{8} + \tfrac{1}{8})] = 77.6 \text{ kips} > 66.7 \quad \text{OK}$$

Check for prying action:

$$Q = F\left[\frac{100bd_b^2 - 18wt_f^2}{70ad_b^2 + 21wt_f^2}\right] \qquad \text{[Eq.(6.27)]}$$

where Q = prying force per fastener
   F = externally applied load per fastener = $\frac{66.7}{4}$ = 16.7 kips
   w = length of flange tributary to each bolt = 4.0 in.
   d_b = nominal bolt diameter = 0.875 in.
   a = distance from fastener line to edge of flange, not to exceed
      $(8.75 - 4.0)/2 = 4.53$, or $2t_f = 2 \times 0.985 = 1.97$ in., $a = 1.97$ in.

[See Eq. (6.27) and AISCM, page 4-81.]
Referring to sketch (b),

$$b = 2.0 - \frac{0.588}{2} - \frac{1}{16} = 1.64 \text{ in.}$$

$$Q = 16.7\left[\frac{(100)(1.64)(0.875^2) - (18)(4)(0.985^2)}{(70)(1.97)(0.875^2) + (21)(4)(0.985^2)}\right] = 4.97 \text{ kips}$$

total bolt tension = $16.7 + 4.97 = 21.67$ kips < 24.05 OK

Check tee flange thickness for added moment due to prying action [Eqs. (6.30) and (6.31). See also AISCM, page 4-82]:

$$M_2 = 4.97 \times 1.97 = 9.79 \text{ kip-in.} \qquad \text{[Eq.(6.30)]}$$

$$M_1 = (21.67 \times 1.64) - (4.97 \times 3.61) = 17.60 \text{ kip-in.} \qquad \text{(governs) [Eq.(6.31)]}$$

Check bending stress in tee flange:

$$f_b = \frac{6 \times 17.60}{4 \times 0.985^2} = 27.2 \text{ ksi} \approx 27 \quad \text{OK}$$

B. Shear connector (use $\tfrac{7}{8}$-in. bolts in friction-type connection)

(1) Web angles to beam:
   Allowable shear strength in double shear:

$$2F_v A_b = 2 \times 15 \times 0.601 = 18.04 \text{ kips/bolt}$$

$$\text{no. reqd} = \frac{54}{18.04} = 2.99$$

Use three high-strength bolts of $\tfrac{7}{8}$-in. diameter to join web angles to beam.

(2) Web angles to column:
Allowable shear strength in single shear is 9.02 kips.

$$\text{no. reqd} = \frac{V}{\text{allowable shear strength}} = \frac{54}{9.02} = 5.98$$

Use six high-strength bolts of $\frac{7}{8}$-in. diameter to join web angles to column.

(3) Required thickness for web angles:
Try 9-in.-long angles.
Required thickness:

$$t = \frac{V}{bF_v}$$

where $V =$ shear force
  $b =$ length of angle
  $F_v =$ allowable shear stress

$$t = \frac{54/2}{9 \times 14.5} = 0.208 \text{ in.}$$

Use two L $4 \times 3\frac{1}{2} \times \frac{5}{16} \times 9$ in. long for web angles.
Check need of column web stiffeners, using Eq. (6.32) instead of AISCS Formula
1.15-1: t of col. web $= 0.34$ in.

$$t = \frac{66.7}{36[0.588 + 5(1.13)]} = 0.30 < 0.34 \qquad \text{OK}$$

No compression stiffeners are needed, and since spread in tension is greater than in compression (through bolts) it may be assumed that no stiffeners are needed in tension.

6.8 CONCLUDING REMARKS CONCERNING CONNECTIONS

As has been stated, connections can be extremely vital elements in a structure. Only a few standard types have been treated herein, and no textbook can really cover the subject adequately. The complexity and variety of connection details is infinite, and their design can only be partially covered by any specification. In special situations connection design requires the application of engineering judgment and experience to a much greater degree than does the design of simple beams and columns. Consider, for example, Fig. 6.15, showing a complex intersection of columns and diagonal elements, shop welded and preassembled for proper fit up of high-strength bolted field connections.The weldment shown is part of a 70-ton assembly to support a 350-ft TV antenna on top of the north tower of the World Trade Center. The assembly provides the transition from the structural framing of the building to the eight bearing plates for the antenna 12 ft above roof level. It was shop assembled in an inverted position (as pictured) and all field connections were match drilled.

Fig. 6.15 Shop assembly of welded and bolted connection for World Trade Center.(Courtesy Montague-Betts Company, Lynchburg, Virginia.)

PROBLEMS

6.1. Design a butt splice similar to that illustrated in Example 6.2, using $\frac{7}{8}$-in.-diameter A325 bolts in a friction-type connection, for A36 steel. Select plate sizes adequate to supply net section to transmit a load of 120 kips in tension and determine the number of bolts required.

6.2. Assume in Example 6.4 that the horizontal component of force is transmitted to the pin plate and channel web in proportion to their respective thicknesses. Select a suitable fillet-weld size for the attachment of pin plate to channel web. Use E70 electrodes.

6.3. Design a bracket connection similar to that of Example 6.3. Select two angles adequate for a tensile force of 160 kips, deducting for a single row of $\frac{7}{8}$-in.-diameter A490 bolts that connect the angles to the web of the WT. Design the connection between flanges of the tee and column using A490 friction-type bolts. Omit design for prying action. Use A36 steel.

6.4. Rework Example 6.4, changing the 160-kip horizontal force component to 200 kips and the 120-kip vertical force component to 150 kips.

6.5. Redesign the bracket connection, Problem 6.3, using welds made with E70 electrode. Follow the general pattern of Example 6.7 but include selection of angle sizes based on the tensile requirement for the gross section.

6.6. Design the eccentrically loaded bracket shown. Assume that there are two plates, one hidden from view on the far side of the column flanges, so that the 110-kip load is divided equally to the two plates. Use A490 $\frac{7}{8}$-in. friction-type high-strength bolts in single shear. Determine the required number. Refer to Example 6.8, also to AISCM, page 4-60.

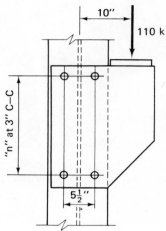

6.7. Design the eccentrically loaded bracket of Problem 6.6 using the dimensions indicated, A36 steel, and E70 electrodes.

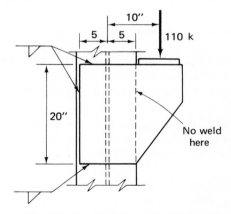

6.8. Same as Problem 6.7, except incline the 110-kip load outward and down at 45° from the horizontal.

6.9. Design of shear-type beam to column connections, utilizing AISCM Tables I through IX, pages 4-12 to 4-59. This problem parallels the five variations covered by Example 6.10.

    (a) Design a bearing-type bolted-end connection for a W 24 × 76 beam, A36 steel, for a reaction of 90 kips. Use $\frac{7}{8}$-in.-diameter A325 bolts with threads not excluded from the shear plane.

(b) Design a friction-type bolted-end connection for a W 30 × 172 beam, A36 steel, for a reaction of 225 kips. Use $\frac{7}{8}$-in.-diameter A325 bolts in a friction-type connection for beam web and outstanding legs.

(c) Design a shop-welded and field-bolted connection, similar to (b), except weld the angles to the beam web and use $\frac{7}{8}$-in.-diameter bolts in a bearing-type connection with threads excluded from the shear plane in the outstanding legs.

(d) Design an all-welded connection, same as (b), except use an E70 weld for connection of both web and outstanding legs of angles.

(e) Design a welded end-plate connection—field bolted—same as (a) except use an E70 welded-end-plate shear connection with $\frac{7}{8}$-in.-diameter A325 bolts in a friction-type connection.

6.10. Design an HS bolted seat and top angle beam-to-column connection for a reaction of 40 kips. The beam is a W 16 × 36 with $F_y = 50$ ksi. Use $\frac{7}{8}$-in. friction-type high-strength bolts.

6.11. Design a welded seat angle using E70 electrodes for the same conditions as in Problem 6.10.

6.12. Design a stiffened seat for a reaction of 65 kips using HS friction-type bolts, a W 16 × 36 beam, and $F_y = 50$ ksi.

6.13. Same as Problem 6.12, except do an all-welded design using E70 electrodes.

6.14. Following the pattern on pages 4-92 to 4-95 of the AISCM, design a shop-welded and field-bolted moment connection for a W 24 × 84 beam framed to a W 12 × 65 column. The design moment is 320 kip-ft and the end reaction is 60 kips. All materials are ASTM A36 steel. Use A325 bearing-type bolts and E70 electrodes.

6.15. Using Example 6.14 as a guide, design a "semirigid" beam-to-column connection, using top and bottom tees and $\frac{7}{8}$-in.-diameter HS A325 bearing-type bolts with threads excluded from the shear plane. A W 27 × 94 beam is connected to a W 14 × 87 column for a bending moment of 120 kip-ft: All steel is type A36. Check for prying action to be included.

Note: A large number of additional problems may be developed simply by requiring a check verification of any of the tabular shear and/or moment values of connections described on pages 4-12 to 4-97 of AISCM.

7

PLATE GIRDERS

7.1 INTRODUCTION

Plate girders are built-up steel beams that require a section modulus greater than any available as a rolled beam. The most common form consists of two heavy flange plates between which is welded a relatively thin web plate. Girder depths range up to 20 ft or more and spans of several hundred feet are not unusual. At points of concentrated load or reaction the girder webs usually must be reinforced by *bearing stiffeners* to distribute the concentrated local forces into the web. *Intermediate* and/or *longitudinal stiffeners* may be added to serve in a quite different role—primarily that of increasing the buckling strength and thereby improving web effectiveness in resisting shear, moment, or combined stresses.

In recent years the economical maximum span of plate girders has greatly increased. This has been made possible in part by the reduction in required web thickness, which results from the use of the tension-field concept that permits utilization of the postbuckling strength of the girder web.

Plate girders are particularly favored for highway bridges; they permit unlimited vision and minimize clearance problems in traffic interchanges and complex multi-level overpasses. Plate girders are also frequently used in various types of buildings and industrial plants to support heavy loads. They are often used, for example, to provide a large space with no interfering columns on a lower floor of a high-rise building, as shown in Fig. 7.1.

Plate girders may be built up with bolts or rivets, as shown in Figs. 7.2(a) and (c), or welded, as shown in Figs. 7.2(b), (d), and (e). In situations where lateral support of the compression flange cannot be provided, box girders [Figs. 7.2(c) and (d)] are especially recommended because of their superior effectiveness against lateral-tor-

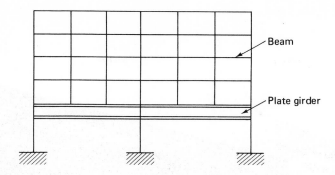

Fig. 7.1   Plate girders in building.

Fig. 7.2   Typical types of plate girders.

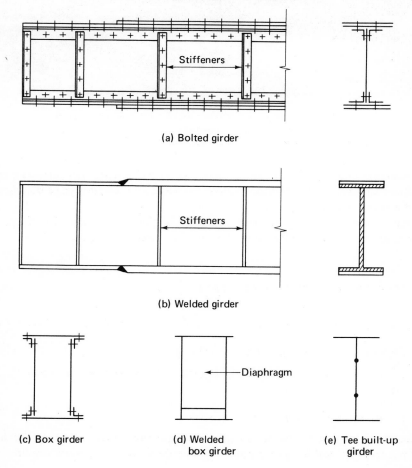

(a) Bolted girder

(b) Welded girder

(c) Box girder   (d) Welded   (e) Tee built-up
                         box girder   girder

Fig. 7.3   Plate girders supporting roof of United Airlines hangar
in San Francisco, consisting of center span with 142 foot cantilevers.
(Courtesy Am. Inst. of Steel Construction.)

sional buckling and in resisting lateral loads. This results from their greater strength
and stiffness in torsion and in bending about the weak axis. The design of box members
in torsion is treated in Chapter 8.

Figure 7.3 shows a long cantilever section of a plate girder being lifted into position for field splicing during the construction of an aircraft hangar.

7.2   SELECTION OF GIRDER WEB PLATE

The selection of the web plate as shown in Fig. 7.4 involves the following steps:

1. Choose a web depth in relation to the span.

2. Choose the minimum thickness in terms of the permissible depth–thickness
   ratio.

1. *Choice of web depth in relation to span*

The depth of girders ranges from one fourteenth to one sixth of the length,
depending on span and load requirements. The shallower girders are required if the

service loads are light; deeper girders will be needed if the loads are heavy or if it is desired to keep deflections at a minimum. It may be desirable to make several preliminary designs with corresponding cost estimates to achieve an optimum depth.

2. Choose web thickness in terms of permissible depth–thickness ratio

The AISCS for plate girders with intermediate stiffeners permits the girder web at allowable loads to go into the postbuckling range to develop tension-field action. After a stiffened thin web panel buckles in shear, it can continue to resist increasing load. The buckled web can resist diagonal tension (left half of Fig. 7.5) much as the diagonals (right half of Fig. 7.5) perform their function as tension members in a Pratt truss. The diagonal web tensions create compressive forces in the intermediate stiffeners, and these vertical stiffeners must be designed to meet this added requirement, thus acting as indicated in a manner analogous to the vertical members of the truss. After the initial buckling of a plate girder web, the girder stiffness decreases rapidly, and the girder deflection may reach a value appreciably greater than predicted by ordinary bending theory.

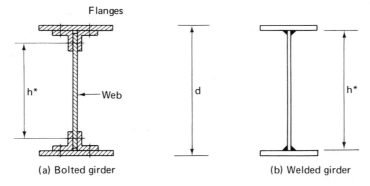

*h as shown to be used as clear depth in calculating web h/t ratios, but h = full web depth in calculating web shear stress

Fig. 7.4 Typical plate girder.

Fig. 7.5 Tension field action in plate girder web analogous to truss with tension diagonals.

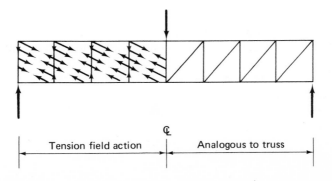

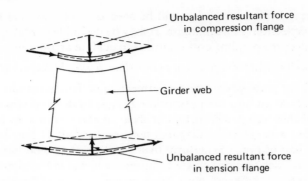

Unbalanced resultant force
in compression flange

Girder web

Unbalanced resultant force
in tension flange

Fig. 7.6 Vertical web compression due to unbalanced force in flanges.

When the postbuckling strength of the web plate is utilized, the criterion for permissible depth–thickness ratio of web is still determined by buckling considerations, but these arise from the fact that under stress the curvature of a plate girder creates vertical compression in the web due to a downward component of flange stress in a curved length of girder on the compression side and an upward component on the tension side, as shown in Fig. 7.6. The vertical buckling strength of the web plate must be sufficient to withstand this squeezing action, and this is taken care of if the depth–thickness ratio of the web meets the AISCS, Supplement No. 2, Sec. 1.10.2 requirement that the ratio of the clear distance between flanges to the web thickness shall not exceed

$$\frac{h}{t} \le \frac{14{,}000}{\sqrt{F_y(F_y + 16.5)}} \tag{7.1}$$

where h = clear distance between flanges (see following note)

   t = web thickness

(See Table 7.1.)

When the foregoing limitation is met, intermediate stiffener spacing as determined by the actual h/t and shear stress f_v is allowed to run as high as $3.0h$. However, if intermediate stiffeners are held to an upper spacing limit of $1.5d$, where d is the girder depth, the maximum permissible h/t ratio is greater; that is,

$$\frac{h}{t} \le \frac{2000}{\sqrt{F_y}} \tag{7.2}$$

(See Table 7.1.)

These upper limits on h/t are listed for various yield stress levels in Table 7.1, where it will be noted that the advantage of the $1.5h$ limit becomes more pronounced as the yield stress increases.

After selection of the web plate, the maximum shear stress $[f_v(\max) = V_{\max}/ht]$ should be determined. If $f_v(\max)$ is less than F_v, by Eq. (1.10-1) (modified for indefinitely large stiffener spacing), and if, at the same time, h/t is less than 260, as well as

Table 7.1

*Maximum ratio of the clear distance between flanges
to web thickness.*

F_y	36	42	45	50	55	60	65	90	100
Eq. (7.1)	322	282	266	243	223	207	192	143	130
Eq. (7.2)	333	309	298	283	270	258	248	211	200

being less than the limit provided by Eq. (7.1), then no intermediate stiffeners are needed at all. Referring to AISCS, Eq. (1.10-1), page 5-28, when a/h is very large, $k = 5.34$, and the allowable shear stress for no stiffeners is given by one of the following equations:

When h/t is greater than $548/\sqrt{F_y}$,

$$F_v = \frac{83{,}150}{(h/t)^2} \tag{7.3}$$

When h/t is less than $548/\sqrt{F_y}$,

$$F_v = \frac{152\sqrt{F_y}}{h/t} \leq 0.4F_y \tag{7.4}$$

For example, what is the allowable shear stress for no stiffeners if $F_y = 36$ ksi and $h/t = 100$? In this case, $548/\sqrt{F_y} = 91$, and Eq. (7.3) applies, giving

$$F_v = \frac{83{,}150}{100^2} = 8.32 \text{ ksi}$$

This value, rounded off to 8.3, could also have been obtained from AISCS, Appendix A, page 5-95, in the extreme right column for "a/h over 3." Equations (7.3) and (7.4) are useful in providing an accurate evaluation of F_v for values of h/t intermediate between those listed in Tables 3-36 to 3-100.

Although not a controlling criterion, the designer at this point has the option of determining whether or not any reduction in allowable bending stress will be required as a result of tension-field action. By AISCS, Sec. 1.10.6, if h/t is less than $760/\sqrt{F_b}$, no reduction is required.

7.3 SELECTION OF GIRDER FLANGES

After the girder web is tentatively selected, the next step is to determine the sizes of the girder flanges. The steps in selecting girder flanges are

1. Preliminary selection.
2. Check width–thickness ratios.

3. Determine reduced allowable bending stress in flanges.

4. Select reduced-sized flanges for use away from maximum moment, and determine location of flange transitions.

(1) *Preliminary flange selection*

As stated in Chapter 3, the required section modulus for a beam is

$$S_x \text{ (reqd)} = \frac{M_x}{F_b} \tag{7.5}$$

and the resisting moment supplied is

$$M_x = F_b S_x \text{ (actual)} \tag{7.6}$$

The bending strength of a plate girder equals the sum of the bending strengths of the girder web and flanges. An approximate evaluation of the girder's bending strength can be obtained as follows:

$$S_x = \frac{I_x}{d/2}$$
$$I_x = I_x \text{ (web)} + I_x \text{ (flanges)}$$

or, approximately

$$I_x \approx \frac{th^3}{12} + 2A_f \left(\frac{h}{2}\right)^2$$

where

$t = $ thickness of girder web
$h = $ depth of girder web
$d = $ depth of girder
$A_f = $ area of one girder flange

Then the section modulus of the girder for $h/d \approx 1$ becomes approximately

$$S_x = \frac{th^2}{6} + A_f h \tag{7.7}$$

Combining Eqs. (7.5), (7.6), and (7.7), the following expression for the required girder flange area is obtained:

$$A_f = \frac{M_x}{F_b h} - \frac{th}{6} \tag{7.8}$$

The first term on the right side of Eq. (7.8) represents the flange area that would be needed to resist the bending moment, M_x, without help from the web. The next term, $th/6$, is the *equivalent* flange area contributed by the girder web. Equation (7.8) provides a tentative trial selection that is subject to later verification by the moment-of-inertia method.

(2) *Trial flange plate selection and b/t check*

After the tentative flange area is determined, a trial flange plate is selected. In order to prevent premature local flange buckling, maximum unsupported width–thickness ratios are imposed, as in column design, by AISCS, Sec. 1.9, as follows:

$$\frac{b}{t_f} < \frac{95}{\sqrt{F_y}} \qquad \text{for unstiffened plates}$$

$$\frac{b}{t_f} < \frac{238}{\sqrt{F_y}} \qquad \text{for stiffened plates}$$

(See AISCS, Appendix A, pages 5-72 to 5-75 for listed values.)

An unstiffened plate is one that is supported along one longitudinal edge and is free along the other, such as the half flange width of an I-shaped girder, or the outstanding width of a bearing stiffener. For the I-shaped girder, b is taken as equal to $b_f/2$, where b_f is the full flange width. The flange of a *box* girder is a stiffened plate, supported along both edges, for which b (in the absence of longitudinal stiffeners) is to be taken as the distance between the nearest lines of fasteners or welds. In light construction, using thin-walled girders, it is sometimes advantageous to exceed the above allowable b/t limits, which is permitted on the basis of reduced allowable stresses for which Appendix C, pages 5-115 to 5-117, provides information. These reductions are not to be confused with the mandatory reduction imposed by tension-field action as discussed in the following section.

(3) *Determine reduced allowable bending stress in flange*

In tension-field design, the web is allowed to buckle, as previously discussed, and the stress on the compression side is no longer proportional to the distance from the neutral axis. As shown in Fig. 7.7, the stress at the extremity of the compression flange is slightly greater than given by the beam formula. To compensate for this in building design, a reduction in allowable flange stress is made according to Sec. 1.10.6 of the AISCS.

When the web depth–thickness ratio

$$\frac{h}{t} \geq \frac{760}{\sqrt{F_b}}$$

Fig. 7.7 Stress distribution after buckling of web.

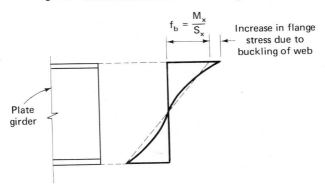

then the maximum stress in the compression flange shall be

$$F'_b \leq F_b\left[1 - 0.0005\frac{A_w}{A_f}\left(\frac{h}{t} - \frac{760}{\sqrt{F_b}}\right)\right],\qquad \text{[AISCS Eq. (1.10-5)]}$$

where F_b = allowable bending stress given in the AISCS, Sec. 1.5.1
            A_w = web area
            A_f = compression flange area

After the trial flange selection is made, the moment of inertia and section modulus are calculated on the basis of actual dimensions and properties. The stress then is calculated and checked against the allowable value.

(4) *Select reduced-sized flanges for use away from maximum moment and determine location of flange area transitions*

For least weight in riveted and bolted girders the flange area should be in proportion to the bending moment. The size of the flange plates can be conveniently reduced in regions where the bending moments have decreased appreciably below the maximum values. However, the AISCS limits the cross-sectional area of cover plates of riveted girders to 70 per cent of the total flange area. For light loadings it is probably reasonable to use a constant cover plate to run the full girder length. In the case of welded girders it is preferable to use a single flange plate, without flange angles or cover plates. The flange plate thickness may be reduced at appropriate intervals, keeping in mind that such reductions should be made only if the saving in the cost of the flange material more than offsets the added expense of introducing the butt welds at thickness transition locations.

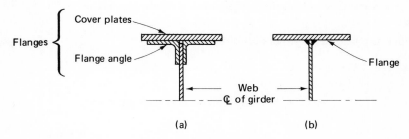

Fig. 7.8 Typical girder flanges.

Cover plates in bolted girders (Fig. 7.8) can be cut off or flange plate thickness in welded girders reduced where the bending moments have dropped appreciably, as shown in Fig. 7.9. The resisting moments of the various girder sections where plates are cut off or reduced in thickness can be obtained by calculating the section modulus of the girder at each location [Eq. (7.5)]. One can locate the thickness transition point (or cut-off point) graphically by plotting horizontal lines that indicate the magnitude of the various resisting moments between transition points. The intersections of the

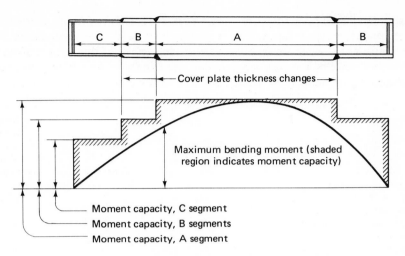

Fig. 7.9   Thickness change locations in plate girder flanges.

moment diagram and the various horizontal lines, as shown in Fig. 7.9, determine the transition points. In the case of a uniform load the transitions can be determined mathematically from the equation for bending moment as follows:

$$\frac{x^2}{(l/2)^2} = \frac{M - M_r}{M} \quad \text{and} \quad \frac{M - M_r}{M} = \frac{A_c}{A}$$

where x = distance from the maximum moment in the girder to the theoretical thickness transition point (or cut-off point)

l = span length of simply supported girder

M = maximum moment in the girder due to uniform load

M_r = resisting moment at theoretical thickness transition point

A_c = area of plate to be cut off or difference of flange area between two flange plates at a transition point

A = flange area plus web equivalent

$= \dfrac{M}{hF_b}$ [see Eq. (7.8)]

Then the theoretical thickness transition point can be found:

$$x = \frac{l}{2} \sqrt{\frac{A_c}{A}} \tag{7.9}$$

If cover plates are used in riveted, bolted, or welded girders, they should be extended beyond the theoretical transition points with enough fasteners or welds to develop the plate stress at the theoretical cut-off points. AISCS, Sec. 1.10.4, provides explicit information regarding extension requirements and details at the ends of cover plates.

In welded design with a single flange plate, transitions in section are best achieved by changing the flange plate thickness (and width, if desired) by joining the ends of the two flange plates with a full-penetration groove butt weld. The weld should have a concave contour that is smoothly tangent with the surface of the thinner plate.

If repeated load is a design requirement, the choice of details may be a critical factor. Referring to AISCS, Appendix B, the following comparisons are tabulated:

	Illustrative Case (AISCS, p. 5-112)	Stress Category (AISCS, Table B-2)	Allowable Stress (500,000 to 2,000,000 repetitions)
Welded girder without stiffeners with no cover plates or splices	(4)	B	17 ksi
Welded groove weld transition (full penetration)	(12)	C	14 ksi
Welded girder with intermediate stiffeners, $f_v > (F_v/2)$	(7)	D	10 ksi
Welded girder with cover plates (stress at cut-off location)	(5)	E	7 ksi

It is readily seen that the use of cut-off welded cover plates may introduce a severe penalty under the conditions stipulated by the foregoing tabulation. If butt joints are used at transition points with full-penetration groove welds, the section may be changed where the reduced section stress would have a maximum stress *range* of no more than 14 ksi—in which case there might not be any penalty due to repeated load.

Flow Chart 7.1 summarizes the steps to be followed in selecting the girder flange sizes.

Flow Chart 7.1

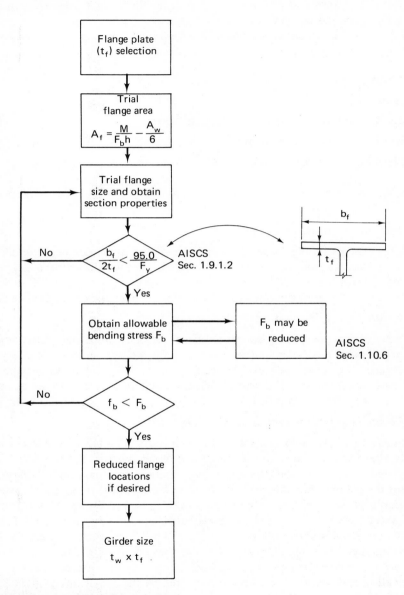

Flange plate
(t_f) selection

Trial
flange area

$$A_f = \frac{M}{F_b h} - \frac{A_w}{6}$$

Trial flange
size and obtain
section properties

No

$$\frac{b_f}{2t_f} < \frac{95.0}{\sqrt{F_y}}$$

AISCS
Sec. 1.9.1.2

b_f

t_f

Yes

Obtain allowable
bending stress F_b

F_b may be
reduced

AISCS
Sec. 1.10.6

No

$f_b < F_b$

Yes

Reduced flange
locations
if desired

Girder size
$t_w \times t_f$

7.4 INTERMEDIATE STIFFENERS

When the shear stress in the web is kept below the allowable value by AISCS Eq. (1.10-1), intermediate stiffeners, if required, serve simply to improve the buckling resistance of the web plate and no tension-field action takes place. For greater shear stress, limited only by AISCS Eq. (1.10-2), the shear strength is assumed to be the sum of the buckled web plate strength and the diagonal tension that is induced in the buckled web. The intermediate stiffeners then play the dual role of improving buckling resistance and acting as compression struts as in truss action.

The design of intermediate stiffeners, when such stiffeners are required, consists of

1. Plotting a maximum shear stress diagram.
2. Locating first stiffeners away from each end.
3. Locating the remaining intermediate stiffeners.
4. Selecting the size of the intermediate stiffeners.
5. Checking the maximum tensile stress in the web.

We consider these now in greater detail.

(1) *Plot maximum shear stress diagram*

Stiffener spacing is a function of shear stress in the particular panel under consideration. It is also a function of the h/t thickness ratio, which has been established at the outset. With the aid of tables in AISCS, Appendix A, the plot of maximum shear stress along the girder permits rapid selection of intermediate stiffener spacing.

(2) *Locate first stiffeners away from each end*

In girder design the spacing between stiffeners at end panels and panels adjacent to those containing large holes shall be such that the web shear stress at end panels does not exceed the value given by AISCS Formula (1.10-1). Thus these panels are designed without any benefit of tension-field action and are assumed to act as anchor panels for the adjacent tension fields.

(3) *Locate remaining intermediate stiffeners*

Referring to Tables 3-36 through 3-100, Appendix A, of AISCS, one enters the table with the given h/t ratio and locates the column wherein the allowable shear stress as listed just exceeds the maximum shear stress in that particular column. The caption for the column then determines the allowable a/h, or stiffener spacing divided by clear depth of web. In these tables the allowable shear stress by AISCS Formula (1.10-1) is used for web design at stress levels below the buckling load, and Formula (1.10-2) at stress levels above the buckling load. Above the buckling load the shear strength is the sum of the buckling strength plus that added by the tension field.

The need for intermediate stiffeners and (if needed) the determination of their proper location and size are complexly interrelated by the provisions of AISCS, Sec. 1.10.2 and 1.10.5.† The sequence of steps in making these determinations is summarized by Flow Chart 7.2.

---

†It is essential at this point that these provisions be studied independently and in conjunction with Flow Chart 7.2, as they will not be repeated in detail in this text.

(4) *Select size of intermediate stiffeners*

According to Sec. 1.10.5.4 of the AISCS, a pair of intermediate stiffeners, or a single intermediate stiffener, shall be so selected as to satisfy

$$I_{st} \geq \left(\frac{h}{50}\right)^4$$

and $$A_{st} \geq \frac{1 - C_v}{2}\left[\frac{a}{h} - \frac{(a/h)^2}{\sqrt{1 + (a/h)^2}}\right] YDht,\qquad \text{[AISCS Formula (1.10-3)]}$$

where C_v, a, h, and t are as defined previously

A_{st} = gross area of a single or pair of intermediate stiffeners, in square inches

I_{st} = moment of inertia of a pair of intermediate stiffeners, or a single intermediate stiffener, with respect to an axis in the plane of the girder web

Y = (yield stress of web steel)/(yield stress of stiffener steel)

D = 1.0 for stiffeners furnished in pairs

= 1.8 for single angle stiffeners

= 2.4 for single plate stiffeners

The moment of inertia limitation is intended to prevent local bending of the web between buckled panels. The area requirement is for the additional purpose of supplying adequate compression strut capacity during tension-field action.

Solutions of the foregoing AISCS Formula (1.10-3) for required area are provided by the italicized listings in Tables 3-36 to 3-100, AISCS, Appendix A, pages 5-95 to 5-103.

It is noted that the intermediate stiffeners may be stopped short of the tension flange at a distance not to exceed four times the web thickness (AISCS, Sec. 1.10.5.4), provided bearing is not needed to transmit a concentrated load or reaction. When single stiffeners are used, they shall be attached to the compression flange, if it consists of a rectangular plate, to resist any uplift tendency due to twist of the flange. Lateral bracing may be attached to the intermediate stiffeners, and these shall then be connected to the compression flange to transmit 1 per cent of the total flange stress, unless the flange is composed only of angles.

(5) *Check maximum bending tensile stress in girder web*

Plate girder webs, which depend upon tension-field action as considered in Formula (1.10-2), shall be so designed that bending tensile stress shall meet the requirement

$$f_{bx} \leq \left(0.825 - 0.375\frac{f_v}{F_v}\right)F_y \leq 0.6F_y,\qquad \text{[AISCS Formula (1.10-7)]}$$

where f_{bx} = bending tensile stress in webs due to moment in the plane of the girder web

f_v = computed average web shear stress (total shear divided by web area)

F_v = allowable shear stress according to Formula (1.10-2)

The use of this formula bypasses the more complex calculation of maximum direct tensile stress, which acts at an angle to the girder axis.

Flow Chart 7.2

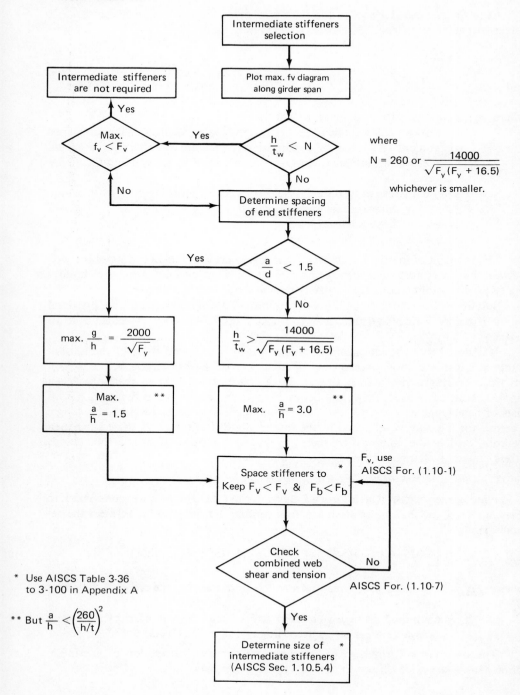

Intermediate stiffeners
selection

Plot max. fv diagram
along girder span

Intermediate stiffeners
are not required

$\dfrac{h}{t_w} < N$

Yes

Yes

Max.
$f_v < F_v$

No

No

where

$N = 260$ or $\dfrac{14000}{\sqrt{F_y (F_y + 16.5)}}$

whichever is smaller.

Determine spacing
of end stiffeners

$\dfrac{a}{d} < 1.5$

Yes

No

max. $\dfrac{g}{h} = \dfrac{2000}{\sqrt{F_y}}$

$\dfrac{h}{t_w} > \dfrac{14000}{\sqrt{F_y (F_y + 16.5)}}$

Max. $\dfrac{a}{h} = 1.5$ **

Max. $\dfrac{a}{h} = 3.0$ **

Space stiffeners to
Keep $F_v < F_v$ & $F_b < F_b$ *

F_v, use
AISCS For. (1.10-1)

Check
combined web
shear and tension

No

Yes

AISCS For. (1.10-7)

Determine size of *
intermediate stiffeners
(AISCS Sec. 1.10.5.4)

* Use AISCS Table 3-36
to 3-100 in Appendix A

** But $\dfrac{a}{h} < \left(\dfrac{260}{h/t}\right)^2$

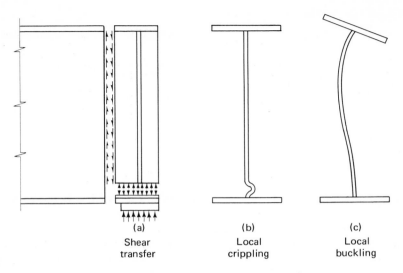

(a)
Shear
transfer

(b)
Local
crippling

(c)
Local
buckling

Fig. 7.10 Support conditions at end of girder.

7.5 BEARING STIFFENERS

Bearing stiffeners serve three interrelated functions, which are illustrated in Fig. 7.10.

1. They transfer local reactive forces to web shear, as illustrated in Fig. 7.10(a).

2. They prevent local crippling in the web immediately adjacent to concentrated reactions or loads. This type of failure is illustrated in Fig. 7.10(b). If no bearing stiffeners are used, the local compressive stress in the web must be checked by AISCS Formula (1.10-8) for interior loads, or (1.10-9) for end loads or reactions. This topic has been treated previously in Sec. 3.9.

3. Finally, bearing stiffeners prevent a more general vertical buckling of the web, of the type illustrated in Fig. 7.10(c). In this connection the allowable average vertical stress components are specified by AISCS Formulas (1.10-10) and (1.10-11), the choice being determined by whether or not the top flange is restrained against rotation. In these formulas the vertical force is assumed to be distributed over a length of web equal to the girder depth or the length of the stiffened panel in which the load is placed—whichever dimension is the lesser.

Bearing stiffeners should have close contact with the flanges adjacent to application points of applied or reactive loads and should extend approximately to the edges of the flanges, as shown in Fig. 7.11. According to AISCS, Sec. 1.10.5.1, the stiffeners are to be designed as columns assuming the column section to comprise the pair of

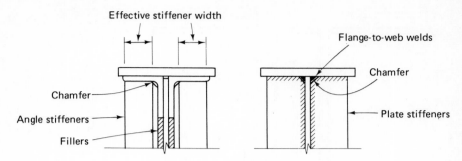

Fig. 7.11 Typical stiffeners.

stiffeners and a centrally located strip of the web whose width is equal to not more than 25 times its thickness at interior stiffeners, or a width equal to not more than 12 times its thickness when the stiffeners are located at the end of the web. The effective length shall be taken as not less than three fourths the length of the stiffeners in computing the slenderness ratio l/r. The stiffeners shall also be checked for local bearing pressure. Only that portion of the stiffeners outside the flange angle fillet or the flange-to-web welds, as shown in Fig. 7.11, shall be considered effective in bearing and the bearing stress shall not exceed the allowable value of $0.90F_y$.

Local detail is important in the transfer of a load concentration into bearing stiffeners. For example, if a very heavy column introduces a load or acts as a support, two pair of stiffeners are desirable so as to introduce the column flange bearing stress directly into the stiffeners without local bending of the girder flanges, as shown in Fig. 7.12.

Fig. 7.12 Bearing stiffeners.

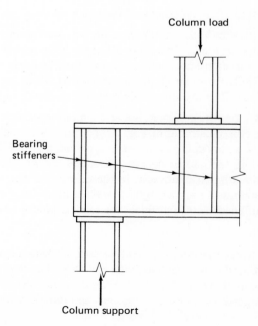

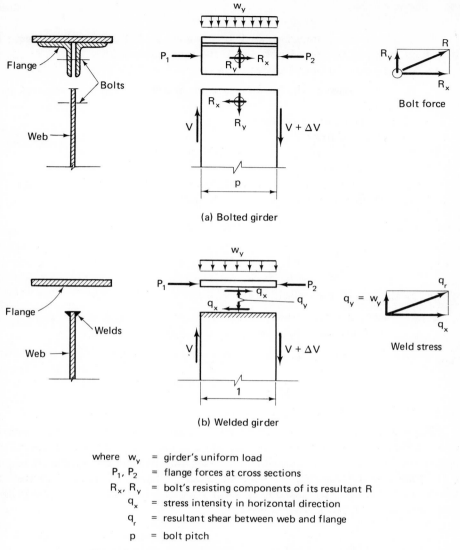

(a) Bolted girder

(b) Welded girder

where w_y = girder's uniform load
P_1, P_2 = flange forces at cross sections
R_x, R_y = bolt's resisting components of its resultant R
q_x = stress intensity in horizontal direction
q_r = resultant shear between web and flange
p = bolt pitch

Fig. 7.13 Stress transmittance from girder flange to web.

7.6 CONNECTIONS OF GIRDER ELEMENTS

(1) *Flange-to-web connection*

Bolts or welds connecting flange to web, or cover plate to flange, shall be designed to resist the horizontal shear resulting from the bending forces on the girder, as shown in Fig. 7.13. The longitudinal spacing of bolts shall not exceed the maximum permissible provided in Sec. 1.18.2.3 for compression elements and in Sec. 1.18.3.1 for tension elements. The flange-to-web connection shall also transmit any direct loads that are applied unless bearing stiffeners are provided.

The flange-to-web connection shall be designed at all locations to transmit the shear force between flange and web due to the moment variation.

(a) In the case of a bolted girder, as shown in Fig. 7.13(a), let

P_1, P_2 = the flange force due to moments M_1, M_2 at any two adjacent cross sections

I = girder moment of inertia

p = pitch of bolts

R_x, R_y = force components of resultant shear force (R) on bolt

V = shear force

$\bar{y}$ = distance between flange centroid and neutral axis of girder section

w_y = vertical applied load per unit length.

Then

$$R_x = P_2 - P_1 = \frac{M_2 - M_1}{I}\bar{y}A_f = \frac{Vp\bar{y}A_f}{I}$$

$$R_y = w_y p$$

$$R = \sqrt{R_x^2 + R_y^2}$$

(b) In the case of the welded girder shown in Fig. 7.13(b), let

q_x, q_y = components of stress resultant loads per unit length of fillet weld

q_r = resultant shear between web and flange

Then

$$q_x = P_2 - P_1 = \frac{M_2 - M_1}{I}\bar{y}A_f = \frac{V\bar{y}A_f}{I}$$

$$q_r = \sqrt{q_x^2 + w_y^2}$$

(2) *Stiffener connections*

The welds, rivets, or bolts that attach bearing stiffeners to the girder web are simply designed to transmit the total reactive or applied load into the web. Similarly, the intermediate stiffeners in a tension-field panel must transfer the vertical component of the total tensile force into the stiffener. Such stiffeners, as required to meet the provisions of AISCS, Sec. 1.10.5.3, at shear stresses permitted by Formula (1.10-2) are to be connected to transmit at least

$$f_{vs} = h\left(\frac{F_y}{340}\right)^{3/2} \text{ kips per linear inch}, \qquad \text{[AISCS Formula (1.10-4)]}$$

where

h = as defined previously

F_y = yield stress of web steel

According to Sec. 1.10.5.4 of the AISCS, the rivets connecting stiffeners to the girder web shall be spaced not more than 12 in. on center. For intermittent fillet welds connecting stiffeners to the girder web, the clear distance between welds shall not be more than 16 times the web thickness nor more than 10 in.

(3) *Web splices*

Splices for girders offer no problem in welded girders, but are expensive and should be avoided whenever possible in riveted or bolted girders. The need for girder

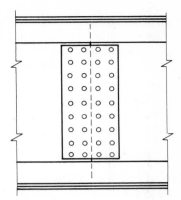

Fig. 7.14 Web splice for riveted or bolted girder.

splices is dictated by erection requirements and shipping limitations. It may be desirable to locate the flange splice and web splice in different locations. The girder web primarily transmits shear force, and therefore web splices are most economically located at places where the shear force is small. Riveted or bolted splices should provide a net section area through the splice plates sufficient to resist the shear force and bending moment carried by the girder web at the splice location. Tests have shown the simple butt splice illustrated in Fig. 7.14 to be quite adequate.

In designing the web splice for moment, the web moment M_w, which is the portion of the total moment (M) carried by the girder web at the splice, can be obtained approximately from the proportion of the equivalent web area to flange, which has been introduced in Sec. 7.3; then

$$M_w = \frac{th/6}{(th/6) + A_f} M$$

After design shears and web moments are determined, the splice design for either bolted or welded connections follows standard procedures for connections, as covered in Chapter 6.

When fillers are needed in the splices, then fillers shall be designed according to Sec. 1.15.6 of the AISCS.

In welded girders, splices present no special problem; complete-penetration groove welds shall be provided to develop the full strength of the smaller spliced section.

7.7 ILLUSTRATIVE EXAMPLE

The AISCM provides illustrative design examples that very adequately cover the use of the specification and the design tables and aids provided in the manual. These design examples, four in all, are described on page 2-107 of the manual. All the girders are welded. Example 1 in the manual and Ex. 7.1 that follows herein are similar, but additional explanation and details of computational steps are provided herein. Example 2 of the AISCM, starting on page 2-114, is of a "hybrid" girder; that is, different yield point steels are used in the flange and web. Tension-field design is not allowed in this case (AISCS, Sec. 1.10.1). The girder in Ex. 2 is more heavily loaded than in the other examples and the determination of the intermediate stiffener spacing, aided by a graphical plot, should be studied carefully in conjunction with Flow Chart 7.2. The AISCM design aids on pages 2-124 to 2-133 will be used in certain of the assigned problems, as specified. Examples 3 and 4 of the AISCM compare two designs under identical load conditions with and without intermediate stiffeners.

Example 7.1

Design a welded plate girder with a simple span of 56 ft to support a uniformly distributed load of 3 kips/ft (girder weight is included) and two concentrated loads of 75 kips located 20 ft from each end. The compression flange is laterally supported only at points of concentrated load. Use A36 steel and the AISCS. It is again noted that the computational detail supplied herein, for explanatory purposes, considerably exceeds what would normally be included on design office calculation sheets.

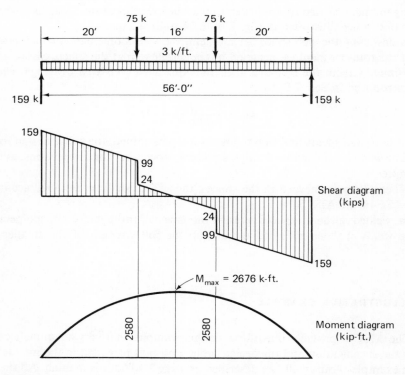

Solution

A. Selection of girder web plate:

(1) Select a girder with a depth of about one eighth of the span length:

$$\frac{l}{8} = \frac{56 \times 12}{8} = 84 \text{ in.}$$

Choose web depth $h = 80$ in.

(2) For no reduction in flange stress (AISCS, Sec. 1.10.6)

$$\frac{h}{t} \le \frac{760}{\sqrt{F_b}} = \frac{760}{\sqrt{22}} = 162$$

Corresponding thickness of web $= h/162 = \frac{80}{162} = 0.494$ in.

(3) Maximum ratio of clear distance between flanges and web thickness (AISCS, Sec. 1.10.2, or AISCM, page 5-74) is

$$\frac{h}{t} = \frac{14,000}{\sqrt{F_y(F_y + 15.5)}} = 322$$

Corresponding minimum thickness of web:

$$t = \frac{h}{322} = \frac{80}{322} = 0.238 \text{ in.}$$

Try web plate $\frac{1}{4} \times 80$; $A_w = 20$ in.^{2}; $h/t = 80/\frac{1}{4} = 320$.
Note: Allowable flange stress will be reduced.

B. Selection of girder flanges:

(1) Preliminary flanges by flange-area method assume $F_b = 20$ ksi [Reduction in flange stress is required; see item A(2)]:

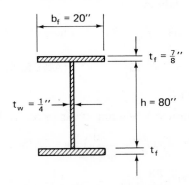

$$A_f = \frac{M}{F_b h} - \frac{A_w}{6} = \frac{2676 \times 12}{20 \times 80} - \frac{20}{6} = 16.8 \text{ in.}^2$$

Try plate $\frac{7}{8} \times 20 = 17.5$ in.^{2} > 16.8 in.^{2} OK

Section properties:

  Moment of inertia:

$$
\begin{aligned}
\text{flanges:} \quad & 2 \times 20 \times \tfrac{7}{8} \times 40.438^2 = 57,233 \\
\text{web:} \quad & \tfrac{1}{12} \times \tfrac{1}{4} \times 80^3 = \underline{10,667} \\
& \qquad\qquad\qquad I_x = \overline{67,900} \text{ in.}^4
\end{aligned}
$$

Section modulus:

$$S_x = \frac{67,900}{40.438} = 1661 \text{ in.}^3$$

Moment of inertia of flange plus $\frac{1}{6}$ web about weak y axis:

$$I_{oy} = \tfrac{1}{12} \times \tfrac{7}{8} \times 20^3 = 583 \text{ in.}^4$$

$$A_f + \frac{A_w}{6} = 17.5 + \tfrac{20}{6} = 20.83 \text{ in.}^2$$

$$r_T = \sqrt{\frac{583}{20.83}} = 5.3 \text{ in.}$$

Note: If dead weight of girder was an estimated part of the design load, it should be rechecked at this point on the basis of web and flange areas. Add 5 per cent to allow for stiffeners and other details of welded girders.

(2) Check width–thickness ratio (AISCS, Sec. 1.9.1.2). The permissible flange width–thickness ratio is

$$\frac{b_f}{2t_f} < \frac{95}{\sqrt{F_y}} = 15.8, \qquad \text{(AISCM, page 5-72)}$$

where

$$\frac{b_f}{2t_f} = \frac{20}{2 \times \tfrac{7}{8}} = 11.4 < 15.8 \qquad \text{OK}$$

(3) Determine allowable bending stress in flanges:
(a) For the 16 ft middle panel,

$$\sqrt{\frac{102 \times 10^3 C_b}{F_y}} = 53 \sqrt{C_b} = 53, \qquad \text{where } C_b = 1$$

$$\frac{l}{r_T} = \frac{16 \times 12}{5.3} = 36.2 < 53 \qquad \text{(AISCS, Sec. 1.5.1.4.6a)}$$

$$F_b = 0.6 F_y = 22 \text{ ksi}$$

The reduced allowable bending stress in the 16 ft panel is

$$F_b' = F_b \left[1 - 0.0005 \frac{A_w}{A_f} \left(\frac{h}{t} - \frac{760}{\sqrt{F_b}} \right) \right], \qquad \text{[AISCS Formula (1.10-5)]}$$

$$= 22 \left[1 - 0.0005 \frac{20}{20.83} \left(320 - \frac{760}{\sqrt{22}} \right) \right] = 19.9 \text{ ksi}$$

Maximum actual bending stress is

$$f_b = \frac{M}{S_x} = \frac{2676 \times 12}{1661} = 19.3 \text{ ksi} < F_b' \qquad \text{OK}$$

(b) For the 20 ft end panels,

$$C_b = 1.75 + 1.05 \frac{M_1}{M_2} + 0.3 \left(\frac{M_1}{M_2} \right)^2, \qquad \text{(AISCS, Sec. 1.5.1.4.6a)}$$

$$= 1.75$$

where
$$\frac{M_1}{M_2} = 0$$

$$\frac{l}{r_T} = \frac{20 \times 12}{5.3} = 45.3 < 53$$

$$F_b = 22 \text{ ksi}$$

$$F'_b = 19.9 \text{ ksi}$$

Maximum actual bending stress is

$$f_b = \frac{M}{S_x} = \frac{2580 \times 12}{1661} = 18.6 \text{ ksi} < F'_b \qquad \text{OK}$$

Use for web, one plate $\frac{1}{4} \times 80$; for flanges, two plates $\frac{7}{8} \times 20$.

C. Intermediate stiffeners:
Stiffeners are not required if $(h/t) < 260$ and $f_v < F_v$;

$$\frac{h}{t} = \frac{80}{\frac{1}{4}} = 320 > 260$$

stiffeners are required and

$$f_v = \frac{V}{A_w} = \frac{159}{20} = 7.95 \text{ ksi}$$

(1) Locate first stiffener away from each end (AISCS, Sec. 1.10.5.3):

$$a \quad \text{or} \quad h \le \frac{348t}{\sqrt{f_v}} = \frac{348 \times \frac{1}{4}}{\sqrt{7.95}} = 30.8 \text{ in.}$$

Place stiffeners 30 in. from each end of girder.
(2) Locate remaining intermediate stiffeners (AISCS Secs. 1.10.5.2 and 1.10.5.3):

$$\frac{a}{h} \le \left(\frac{260}{h/t}\right)^2 = \left(\frac{260}{320}\right)^2 = 0.66, \qquad a \le 0.66 \times 80 = 52.8 \text{ in.}$$

(a) For spacing between end stiffener and concentrated load refer to Table 3-36, AISCS, Appendix A, page 5-95. Enter table on line for $h/t = 320$. Extrapolating beyond the column for $a/h = 0.6$, the allowable shear stress $F_v = 9.7$ ksi, for $a/h = 0.66$. Check shear stress 30 in. from end of girder, at first stiffener.

$$V = 159 - 3 \times 2.5 = 151.5 \text{ kips}$$
$$f_v = \frac{V}{A_w} = \frac{151.5}{20} = 7.58 \text{ ksi} < 9.7 \qquad \text{OK}$$

Thus for the total distance between the first stiffener and the concentrated applied load, it would be satisfactory to use four spaces at $52\frac{1}{2}$ in., very nearly the maximum allowed. However, later consideration of the central region between concentrated loads will require four spaces at 48 in., since three spaces in 16 ft would overrun the allowable. Therefore, for a more balanced design, we shall use five spaces at 42 in. in the end segments.

The following check on allowable shear stress is unnecessary, in view of the availability of Table 3-36, but is included to illustrate the computations that are involved:

$$\frac{a}{h} = \frac{42}{80} = 0.525$$

$$k = 4 + \frac{5.34}{(a/h)^2} = 4 + \frac{5.34}{0.525^2} = 4 + 19.4 = 23.4$$

$$C_v = \frac{45,000k}{F_y(h/t)^2} = \frac{45,000 \times 23.4}{36 \times 320^2} = 0.287 < 0.8$$

$$F_v = \frac{F_y}{2.89}\left[C_v + \frac{1 - C_v}{1.15\sqrt{1 + (a/h)^2}}\right], \qquad \text{[AISCS Formula (1.10-2)]}$$

$$= \frac{36}{2.89}\left(0.287 + \frac{1 - 0.287}{1.15\sqrt{1 + 0.525^2}}\right)$$

$$= \frac{36}{2.89}\left(0.287 + \frac{0.713}{1.15 \times 1.13}\right) = \frac{36}{2.89}(0.287 + 0.55) = 10.4 \text{ ksi}$$

(See AISCS, Table 3-36.)

(b) Spacing in the midspan region between concentrated loads is a total of 16 ft. On the basis of previous calculations, a limit of 52.8 in. would be satisfactory. However, the only option is to use four spaces at 48 in., as three at 64 in. would exceed the limit appreciably. Check maximum bending tensile stress in girder web (Sec. 1.10.7):

$$F_{bx} = \left(0.825 - 0.375\frac{f_v}{F_v}\right)F_y \leq 0.6F_y$$

$$= \left(0.825 - 0.375\frac{4.95}{10}\right)F_y = 0.64F_y$$

Use $F_{bx} = 0.6F_y = 22$ ksi > bending stress in web.
Use a spacing of five at 42 in. between end stiffener and concentrated load, and a spacing of four at 48 in. in the center region.

(3) Select size of intermediate stiffeners (AISCS, Sec. 1.10.5.4):

Stiffener area:

$$A_{st} = \frac{1 - C_v}{2}\left[\frac{a}{h} - \frac{(a/h)^2}{\sqrt{1 + (a/h)^2}}\right]YDht, \qquad \text{[AISCS Formula (1.10-3)]}$$

Or use AISCS, Table 3-36, for $h/t = 320$ and $a/h = 0.60$; then

$$A_{st} = 0.112ht = 0.112 \times 80 \times \tfrac{1}{4} = 2.24 \text{ in.}^2$$

Actual area required:

$$\frac{f_v}{F_v}A_{st} = \frac{7.58}{10.4}2.24 = 1.63 \text{ in.}^2$$

Try two plates $\tfrac{1}{4} \times 4 = 2.0$ in.^{2} > 1.63.   OK

(a) Check width–thickness ratio (AISCS, Sec. 1.9.1.2):

$$\frac{b'}{t} = \frac{4}{\frac{1}{4}} = 16 \approx 15.8 \qquad \text{OK} \qquad \text{(page 5-72)}$$

(b) Check moment of inertia (AISCS, Sec. 1.10.5.4):

$$I_{st} \geq \left(\frac{h}{50}\right)^4 = \left(\frac{80}{50}\right)^4 = 6.55 \text{ in.}^4 \qquad \text{required}$$

$$I_{st} = \tfrac{1}{12} 0.25 (2 \times 4 + 0.25)^3 = 11.7 \text{ in.}^4 \text{ furnished} > 6.55 \qquad \text{OK}$$

(c) Length required (AISCS, Sec. 1.10.5.4):

$$h - 4t = 80 - 4 \times \tfrac{1}{4} = 79 \text{ in.}$$

Use for intermediate stiffeners two plates $\frac{1}{4} \times 4 \times 6$ ft 7 in. bearing on compression flange of girder.

D. Bearing stiffeners (AISCS, Sec. 1.10.5.1):

Under concentrated loads and at end of girder, design for end reaction. Since bearing stiffeners must extend approximately to edges of flange plates, try two $\frac{1}{2} \times 8$ plates.

(1) Check width–thickness ratio (AISCS, Sec. 1.9.1.2):

$$\frac{b}{t} = \frac{8}{0.5} = 16 \approx 15.8 \qquad \text{OK}$$

(2) Check web crippling (AISCS, Sec. 1.10.10):

(a) Resulting from a distributed load of 3 kips/ft,

$$f_a \leq \left[5.5 + \frac{4}{(a/h)^2}\right]\frac{10,000}{(h/t)^2}$$

where

$$f_a = \frac{P}{bt} = \frac{3}{12 \times \frac{1}{4}} = 1.0 \text{ ksi}$$

Right term:

$$\left(5.5 + \frac{4}{0.6^2}\right)\frac{10,000}{320^2} = 1.63 \text{ ksi} > 1.0 \qquad \text{OK}$$

(b) Resulting from a concentrated load of 159 kips at ends,

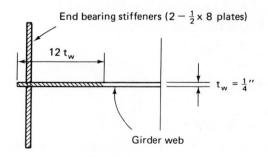

End bearing stiffeners ($2 - \frac{1}{2} \times 8$ plates)

$12\,t_w$

$t_w = \frac{1}{4}''$

Girder web

$$I = \tfrac{1}{12} \times 0.5 \times 16.25^3 = 179 \text{ in.}^4$$
$$A_{\text{eff}} = 2 \times 0.5 \times 8 + 12 \times 0.25^2 = 8.75 \text{ in.}^2$$
$$r = \sqrt{\frac{I}{A}} = \sqrt{\frac{179}{8.75}} = 4.52 \text{ in.}$$
$$l = \tfrac{3}{4}h = \tfrac{3}{4} \times 80 = 60 \text{ in.}$$
$$\frac{l}{r} = \frac{60}{4.52} = 13.3$$

From AISCS, Table 1-36, for $Kl/r = 13.3$, then

$$F_a = 20.9 \text{ ksi}$$
$$f_a = \frac{P}{A_{\text{eff}}} = \frac{159}{8.75} = 18.2 \text{ ksi} < F_a \qquad \text{OK}$$

Use two plates $\tfrac{1}{2} \times 8 \times 6$ ft 8 in. as bearing stiffeners, with close fit against flanges that receive the concentrated loads.

Use same size bearing stiffeners under concentrated loads.†

E. Connections (use E70XX electrode):
(1) Flange to web connection:
Horizontal shear:

$$q_x = \frac{V\bar{y}A_f}{I} = \frac{159 \times 40.44 \times 17.5}{67,900} = 1.66 \text{ kips/in.}$$

Vertical shear:

$$w_y = \tfrac{3}{12} = 0.25 \text{ kips/in.}$$

Resultant shear:

$$q_r = (q_x^2 + q_y^2)^{1/2}$$
$$= (1.66^2 + 0.25^2)^{1/2} = 1.68 \text{ kips/in.}$$

Required weld size:

$$\frac{q_r/2}{0.707F_v} = \frac{1.68/2}{0.707 \times 21} = 0.057 \text{ in.}$$

Minimum weld size for $\tfrac{7}{8}$-in. flange plate:

$$w_{\min} = \tfrac{5}{16} \text{ in.}, \qquad \text{AISCS, Table 1.17.5}$$

Allowable strength for $\tfrac{5}{16}$-in. welds:

$$q_a = 0.707wF_v$$
$$= 0.707 \times \tfrac{5}{16} \times 21 = 4.64 \text{ kips/in.}$$

Minimum length of fillet welds (AISCS, Sec. 1.17.7):

$$l_{\min} = 4w = 4 \times \tfrac{5}{16} = 1\tfrac{1}{4} \text{ in.}$$

Maximum spacing (AISCS, Sec. 1.18.3.1):

$$a_{\max} = 24t \quad \text{or} \quad 12 \text{ in.}, \qquad \text{whichever is smaller}$$
$$= 24 \times \tfrac{5}{16} = 7.5 \text{ in.}$$

†*Note:* For interior bearing, $25t$ may be used in determining the effective web area under concentrated loads at interior panels according to Sec. 1.10.5.1 of the AISCS.

Try $\frac{5}{16}$-in. welds, $1\frac{1}{2}$ in. long; then the required spacing is

$$a = \frac{2lq_a}{q_r} = \frac{2 \times 1.5 \times 4.64}{1.68} = 8.3 \text{ in.}$$

Use $\frac{5}{16} \times 1\frac{1}{2}$ in. welds, 6 in. on centers.
(2) Stiffener connections:
(a) Welds between web and intermediate stiffeners:
Transferred shear (AISCS, Sec. 1.10.5.4):

$$f_{vs} = h\left(\frac{F_y}{340}\right)^{3/2} = 0.034h = 0.034 \times 80 = 2.72 \text{ kips/in.}$$

Required weld size:

$$\frac{f_{vs}/2}{0.707F_v} = \frac{2.72/2}{0.707 \times 21} = 0.0915 \text{ in.}$$

Minimum weld size for $\frac{1}{4}$-in. plate is $\frac{1}{8}$ in. (AISCS, Sec. 1.17.5)
Allowable strength for $\frac{1}{4}$-in. welds:

$$q_a = 0.707 \times \frac{1}{4} \times 21 = 3.71 \text{ kips/in.}$$

Minimum length of fillet welds:

$$4w = 4 \times \frac{1}{4} = 1 \text{ in.}$$

Maximum spacing:

$$24t = 24 \times \frac{1}{4} = 6 \text{ in.} < 12 \text{ in.} \qquad \text{OK}$$

Try $\frac{1}{4}$-in. welds, $1\frac{1}{2}$ in. in length. Required spacing:

$$a = \frac{2lq_a}{f_{vs}} = \frac{2 \times 1.5 \times 3.71}{2.72} = 4.1 \text{ in.}$$

Use $\frac{1}{4} \times 1\frac{1}{2}$ in. welds, 4 in. on centers.
(b) Welds between bearing stiffeners and web:

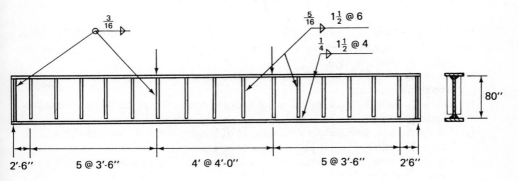

Web: 1 plate $\frac{1}{4}$ x 80

Flanges: 2 plates $\frac{7}{8}$ x 20

Intermediate stiffeners : 2 plates $\frac{1}{4}$ x 4 x 6'-7"

Bearing stiffeners : 2 plates $\frac{1}{2}$ x 8 x 6'-8"

Try minimum weld size, $\frac{3}{16}$-in. continuous welds both sides; then $\frac{3}{16}$-in. weld strength is

$$q_a = 0.707 \times \tfrac{3}{16} \times 21 = 2.78 \text{ kips/in.}$$

Total weld strength:

$$2q_a h = 2 \times 2.78 \times 80 = 444 \text{ kips} > 159 \text{ kips} \qquad \text{OK}$$

PROBLEMS†

7.1. For a welded girder cross section made up of single $\frac{7}{8} \times 22$ flange plates, top and bottom, connected by a $\frac{3}{8} \times 80$ web plate, determine
   (a) The approximate resisting moment based on the flange area method.
   (b) The resisting moment by the moment of inertia method.
   (c) The required stiffener spacing at a location where the shear is 300 kips.

7.2. The girder cross section shown is made up of a C18 × 42.7 channel attached to the top flange of a W30 × 108 section by means of friction-type A325 high-strength bolts of $\frac{3}{4}$-in. diameter. What bolt pitch is required at a location where the shear on the girder is 210 kips?

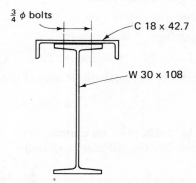

7.3. On the basis of the structural details illustrated on the facing page, using E60 welding electrodes, determine the permissible end reaction R (same as end shear V), in each of the following five different ways:
   (a) On the basis of the fillet weld size and weld spacing at the end of the girder, attaching web to flange plates.
   (b) On the basis of the local contact compressive bearing stress at the bottom ends of the bearing stiffeners.
   (c) On the basis of the size and arrangement of bearing stiffeners. (Assume that welds between bearing stiffeners and web are adequate.)
   (d) On the basis of the first space (40 in.) from the bearing stiffener to the first inter-mediate stiffener.
   (e) On the basis of the second space (60 in.) between the first and second intermediate stiffeners. (The shear used in the design of this space is assumed to be 20 kips less than the end reaction, that is, $V = R - 20$.)

† Assume use of A36 steel in all problems.

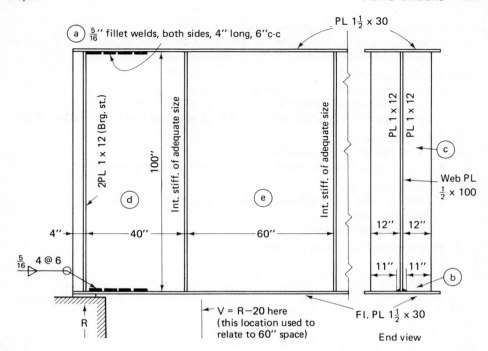

7.4. Design an all-welded plate girder for the following conditions. Include a general design drawing and detailed sketches of such portions as are judged needed to provide complete information. Span, 140 ft. Simple supports. A36 steel. Uniform live load of 2 kips/ft. Concentrated load of 840 kips, 28 ft from left support. Assume adequate lateral support. Web plates are to be as thin as possible, in commercially available thicknesses, to carry the maximum shear. Web plate thickness may be varied, if desired, to suit the high shear area at one end.

Note: This is a rather lengthy assignment and is suitable as a term project. For classroom use it is suggested that groups of three or four students each be assigned arbitrary web plate depths, between 110 and 160 in., varying in increments of 10 in. A plot of girder weight versus web plate depth can then be made as an exercise in the search for the most economical proportion.

The following is a summary† of steps that may be followed in carrying out this assignment:

1. Select web plate.
   a. Choose clear depth in relation to span.
   b. Choose thickness. (1.10.2)
   c. Check for shear. (1.10.5.2)
   d. Check maximum tension stress in web. (1.10.7)

2. Flange plates.
   a. Preliminary selection by flange-area method.
   b. Determine reduced allowable stress. (1.10.6)

† Parenthetical references are AISCS sections.

    c. Check stresses due to bending by Mc/I.

    d. Select reduced size flanges for use away from maximum moment and determine location of flange transitions. (Check flange width–thickness ratios.) (1.9.2.2)

3. Intermediate stiffeners.
    a. Locate first stiffener away from each end. (1.10.53)
    b. Locate remaining intermediate stiffeners. (1.10.5.2–1.10.5.3)
    c. Select size of intermediate stiffeners. (Check area and I requirements.) (1.10.5.4)

4. Bearing stiffeners.
    a. Design for maximum reaction. (1.10.5.1)
    b. (For assigned problem, assume other bearing stiffeners to be identical.)

5. Design welds.
    a. Stress transfer for intermediate stiffeners. (1.10.5.1)
    b. Bearing stiffeners.
    c. Web-to-flange shear transfer.

6. Weight takeoff for complete girder.

8

SPECIAL TOPICS

IN BEAM DESIGN

8.1 INTRODUCTION

Chapters 3 and 7 included beam and plate girder design problems for which specification coverage is adequate, including the usual problems arising from lack of lateral support of rolled shapes having an axis of symmetry in the plane of the loads. We now consider special problems that arise in beam design as a result of lack of lateral support, in combination with loads that are not in a plane of symmetry, such as is usually the case when the channel, angle, or other unsymmetrical section is used.† Special attention is given to the problem of combined bending and torsion, and a simplified procedure is presented for W beam shapes.

Composite design is mentioned briefly, but for detailed coverage the reader is referred to the AISCM, which provides a very adequate presentation of this topic.

Continuous beam design examples for short-span building design applications include both the traditional allowable-stress approach and an introduction to the plastic-design method that was introduced in general terms in Chapter 3.

8.2 TORSION

If torsion in a structural member is a major part of the load system, either a cylindrical or box tube should be used if possible. The cylindrical tube utilizes material

†Refer to Sec. 3.1 for a review of support conditions required to permit use of simple bending theory in design.

in the most effective way possible for resistance to torsion. Box shapes are a close second. The tube in Fig. 8.1 is loaded in pure torsion, as is the drive shaft of an automobile or the propellor shaft of a ship under its primary load. In Fig. 8.1, the torsional moment, M_t, is equal to Pa and the end twists through a total angle ϕ. For a tubular member under uniform pure torsion the angle of twist per unit length is constant:

$$\theta = \frac{\phi}{l} \tag{8.1}$$

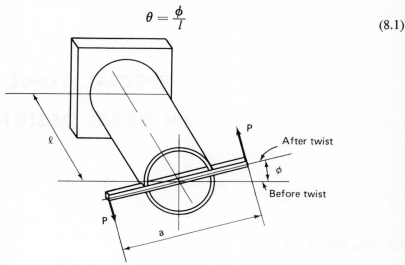

Fig. 8.1 Hollow cylinder in pure torsion.

The stress f_v in a torqued tube is "pure shear," as is indicated in Fig. 8.2. In a thin-walled tube the shear stress can be assumed constant through the wall thickness t, and each unit distance around the circumference exerts a tangential force equal to tf_v. The twisting moment about the central axis of the cylinder, at O, of each unit length of tangential shear force is $tf_v r$, where r is the mean radius of the cylinder. Summing up the contributions of each unit length of circumference, the total torsional moment is equal to $tf_v r$ multiplied by the circumferential distance; hence

$$M_t = 2\pi t f_v r^2 \tag{8.2}$$

Fig. 8.2 Section through a hollow cylinder in pure torsion.

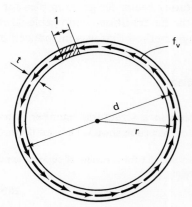

The mean radius of the tube encompasses an area equal to

$$A_o = \pi r^2$$

Hence an alternative expression to Eq. (8.2) can be written

$$M_t = 2A_o t f_v \tag{8.3}$$

The form of Eq. (8.3) is useful in that it applies to square and rectangular box tubes as well as cylindrical. If the wall thickness of a box section varies, Eq. (8.3) still applies, but t should be taken as the thickness of the thinnest plate segment, as this will be the most highly stressed and will determine the allowable torsional moment.

When a box section is twisted, plane sections before twist remain plane—or very nearly so—and their contribution to the torsional resistance is in proportion to their distance from the center of twist. When an "open" section, such as a wide-flange shape, is twisted, elements not centered on the axis of twist *tilt* or *warp*, and unless such warping is in some way restrained the torsional resistance of each component is independent of its location in the cross section. Thus "closed" or box members are many times more rigid than "open" sections of the same general dimensions and weight per unit length. The torsional rigidity of a member is measured by the torsion constant J of the cross section, just as the bending rigidity is measured by the moment of inertia I. For a member under uniform torsion with no section restrained against warping, the general relationship between torsional moment and angle of twist per unit length is

$$M_t = JG\theta \tag{8.4}$$

For a circular cross section, solid or hollow, J is equal to the polar moment of inertia; for any noncircular section it is always less than the polar moment of inertia. For a closed box section of *any* shape, enclosing only one internal cell,

$$J = \frac{4A_o^2}{\sum\limits_{i=1}^{n} \dfrac{s_i}{t_i}} \tag{8.5}$$

The denominator in Eq. (8.5) is the summation of the length–thickness ratios of all the n component parts of the tube around the periphery of the cross section. Thus for a thin-walled hollow cylinder, $A_o = \pi d^2/4$ and $\sum (s_i/t_i) = \pi d/t$, and for the circular pipe or tube

$$J = \frac{\pi d^3 t}{4} \tag{8.6}$$

For the box beam shown in Fig. 8.3, $\sum(s_i/t_i) = 2(h/t_w + b/t_f)$, and by Eq. (8.5),

$$J = \frac{2b^2h^2}{(h/t_w) + (b/t_f)} \tag{8.7}$$

The torsion constant of a solid rectangular bar section, several times wider than its thickness, is approximately

$$J = \tfrac{1}{3}bt^3 \tag{8.8}$$

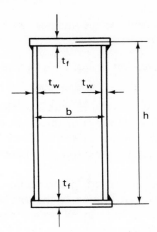

Fig. 8.3 Box section nomenclature.

The torsion constant of a structural shape, such as a wide-flange beam or angle, is simply approximated by summing Eq. (8.8) for the various component rectangular parts. More accurately, the AISCM lists J for standard shapes.

The maximum shear stress in a structural shape of open section under uniform torsion is

$$f_v = \frac{M_t t}{J} \tag{8.9}$$

In Eq. (8.9) the maximum shear stress is obviously located where the thickness t is greatest.

When flange warping is unrestrained, the distribution of shear stress in a wide-flange shape under uniform torsion is as shown in Fig. 8.4(a). When warping is restrained, the shear stress is nearly constant through the flange thickness, as shown in Fig. 8.4(b). The flanges are then stressed in shear as they would be in a rectangular beam. Warping restraint can be accomplished most effectively in a section by "boxing" a length of the wide-flange shape, as shown in Fig. 8.4(c), with added stiffeners that should be as long as the beam is deep.

At sections away from a restrained location, the stress distribution is a combination of that shown in Figs. 8.4(a) and (b), approaching that of Fig. 8.4(a) as the

Fig. 8.4 Torsional shear stress distributions at unrestrained and restrained locations in a wide flange beam.

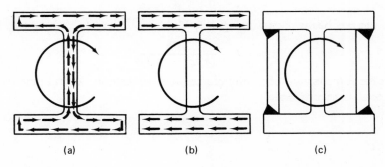

(a)   (b)   (c)

distance from the restrained location increases. A *warping constant* C_w is also tabulated in the AISCM. Another torsion constant, the *torsion bending constant*, a, gives a rough approximation of the distance along a beam away from a restrained location that is required for the effect of restraint to be dissipated and the condition of Fig. 8.4(a) to be approached. The torsion bending constant may be calculated readily from the listed values of J and C_w in the AISCM.

$$a = 1.61\sqrt{\frac{C_w}{J}} \qquad (8.10)$$

The dimensionless parameter l/a will be used in the next section in a simplified presentation of combined bending and torsion with comparative design examples involving both box and open sections.

8.3 COMBINED BENDING AND TORSION

In the structural design of buildings it is desirable and usually possible to avoid the complications of torsion by locating members so that they are required to transmit only shears, moments, and direct stresses, alone or in combination. But if torsion cannot be avoided in a laterally loaded beam, the problem involves *combined bending and torsion*, and the rectangular box section should be used if feasible. The AISCM provides dimensions and properties of available pipe and both square and rectangular box tubing (pages 1-102 to 1-106).

In designing a box beam for combined bending and torsion, the preliminary selection may be made for bending moment alone, with a subsequent check on the combined shear stress due to both bending and torison. More often than not the preliminary selection will be adequate. Care should be taken where local concentration of torque loads is introduced to guard against local distortion of the cross section. External stiffeners or internal diaphragms may be needed. If the framing is conducive to the use of a triangular closed box section, cross-sectional distortion is automatically eliminated.

For very short and stubby members, a W section may be suitable in combined bending and torsion. If the length l of a cantilever beam is less than $0.5a$ [Eq. (8.10)], the individual flanges may be assumed to take all the torsional moment as if they were individual cantilever beams, loaded in opposite directions. The cantilever beam shown in Fig. 8.5(a) is to be designed for a cantilever bending moment equal to Pl combined with a torsional moment equal to Pe. These may be considered separately for the loads shown in Fig. 8.5(b) and, for the relatively short beam with l/a less than 0.5, as shown in Fig. 8.5(c). If the ratio $C_n = S_x/S_y$ can be estimated (see discussion in Sec. 3.8 on biaxial bending), a preliminary estimate of the required section modulus may be made. The section modulus of an individual flange of a W section is approximately $S_y/2$. Thus, for bending and torsional moments

$$M_x = Pl \quad \text{and} \quad M_f = \frac{Pel}{h}$$

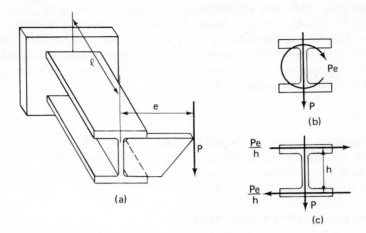

Fig. 8.5 Very short W beam section assumed to resist torsion by flange shear forces.

the required section modulus is

$$S_x = \frac{M_x + 2C_n M_f}{F_b} \tag{8.11}$$

For cantilever loads producing bending about the weak bending axis a more direct selection may be made, the required section modulus about yy being

$$S_y = \frac{M_y + 2M_f}{F_b} \tag{8.12}$$

Simplified formulas for the stress in very long W beams can be written; but unless the maximum permissible angle of twist is very large a tubular member will be desirable. Tables 8.1 and 8.2 tabulate simple approximate formulas for maximum flange moment and maximum angle of twist for the six cases of combined bending and torsion that are most commonly encountered. The formulas include simple approximations for short and long beams and interpolation formulas for intermediate-length beams. The principal advantage of these formulas over the exact ones is the fact that they may be used for direct design checks without reference to tables of hyperbolic or exponential functions. Formulas for the maximum shear stress due to torsion are not tabulated.

For short beams in the minimum l/a category, the shear stress will be mostly of the type shown in Fig. 8.4(b). Neglecting the shear stress of the type in Fig. 8.4(a), the maximum shear stress may be assumed to be 1.5 times the average at the location along the beam where shear due to torsional flange bending is greatest. To calculate the flange shear due to torsional flange bending, each individual flange may be assumed to act as a beam, loaded as shown in Fig. 8.5(c), with load magnitudes equal to the total beam load times e/h. The load distributions and beam end conditions for the individual flanges in torsional bending will be identical with those shown in Tables

Table 8.1

Approximate formulas for flange moment and total
twist angle. Three concentrated load cases, 1, 2, and 3.

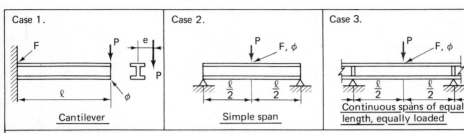

Case 1.	Case 2.	Case 3.
Cantilever	Simple span	Continuous spans of equal length, equally loaded

Maximum lateral bending moment in each flange at location F.

ℓ/a less than 0.5:	ℓ/a less than 1.0 :	ℓ/a less than 2.0 :
$M_f = \dfrac{Pe\ell}{h}$	$M_f = \dfrac{Pe\ell}{4h}$	$M_f = \dfrac{Pe\ell}{8h}$
ℓ/a more than 2.0 :	ℓ/a more than 4.0 :	ℓ/a more than 8.0
$M_f = \dfrac{Pea}{h}$	$M_f = \dfrac{Pea}{4h}$	$M_f = \dfrac{Pea}{8h}$

Maximum total angle of twist at end of a cantilever or a midspan of other beams, at location ϕ.

ℓ/a less than 0.5 :	ℓ/a less than 1.0 :	ℓ/a less than 2.0 :
$\phi_t = 0.32 \dfrac{Pea}{JG} \left(\dfrac{\ell}{a}\right)^3$	$\phi_t = 0.32 \dfrac{Pea}{JG} \left(\dfrac{\ell}{2a}\right)^3$	$\phi_t = 0.64 \dfrac{Pea}{JG} \left(\dfrac{\ell}{2a}\right)^3$
ℓ/a more than 2.0 :	ℓ/a more than 4.0 :	ℓ/a more than 8.0 :
$\phi_t = \dfrac{Pe}{JG} (\ell - a)$	$\phi_t = \dfrac{Pe}{JG} \left(\dfrac{\ell}{2} - a\right)$	$\phi_t = \dfrac{2Pe}{JG} \left(\dfrac{\ell}{4} - a\right)$

Interpolation formulas for bending moment in each flange at F and maximum total twist angle at location ϕ.*

Case 1.   Cantilever. ℓ/a more than 0.5 and less than 2.0 :

$$M_f = \frac{Pea}{h} \left[0.05 + 0.94 \left(\frac{\ell}{a}\right) - 0.24 \left(\frac{\ell}{a}\right)^2\right], \phi_t = \frac{Pea}{JG} \left[-0.029 + 0.266 \left(\frac{\ell}{a}\right)^2\right]$$

Case 2.   Simple Span. ℓ/a more than 1.0 and less than 4.0 :

$$M_f = \frac{Pea}{2h} \left[0.05 + 0.94 \left(\frac{\ell}{2a}\right) - 0.24 \left(\frac{\ell}{2a}\right)^2\right], \phi_t = \frac{Pea}{JG} \left[-0.029 + 0.266 \left(\frac{\ell}{2a}\right)^2\right]$$

Case 3.   Continuous spans of equal length, equally loaded.
        ℓ/a more than 2.0 and less than 8.0 :

$$M_f = \frac{Pea}{2h} \left[0.05 + 0.94 \left(\frac{\ell}{4a}\right) - 0.24 \left(\frac{\ell}{4a}\right)^2\right], \phi_t = \frac{2Pea}{JG} \left[-0.029 + 0.266 \left(\frac{\ell}{4a}\right)^2\right]$$

*Note.   At the extreme range of application the interpolation formulas are more accurate than the formulas given for locations outside of this range, yielding values slightly less than the others, which err slightly on the side of conservative design estimates.

Table 8.2

*Approximate formulas for flange moment and total
twist angle. Three uniform load cases, 4, 5 and 6.*

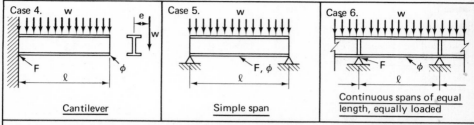

Case 4.	Case 5.	Case 6.
Cantilever	Simple span	Continuous spans of equal length, equally loaded

Maximum lateral bending moment in each flange at location F.

ℓ/a less than 0.5 :	ℓ/a less than 1.0 :	ℓ/a less than 2.0 :
$M_f = \dfrac{w\ell^2 e}{2h}$	$M_f = \dfrac{w\ell^2 e}{8h}$	$M_f = \dfrac{w\ell^2 e}{12h}$
ℓ/a more than 3.0 :	ℓ/a more than 6.0 :	ℓ/a more than 8.0 :
$M_f = \dfrac{w\ell ea}{h}\left(1 - \dfrac{a}{\ell}\right)$	$M_f = \dfrac{wea^2}{h}$	$M_f = \dfrac{w\ell ea}{h}\left(\dfrac{1}{2} - \dfrac{a}{\ell}\right)$

**Maximum total angle of twist at end of a cantilever or at midspan
of other beams, at locations ϕ.**

ℓ/a less than 0.5 :	ℓ/a less than 1.0 :	ℓ/a less than 2.0 :
$\phi_t = 0.114\,\dfrac{w\ell ea}{JG}\left(\dfrac{\ell}{a}\right)^3$	$\phi_t = 0.094\,\dfrac{w\ell ea}{JG}\left(\dfrac{\ell}{2a}\right)^3$	$\phi_t = 0.151\,\dfrac{w\ell ea}{JG}\left(\dfrac{\ell}{4a}\right)^3$
ℓ/a more than 3.0 :	ℓ/a more than 6.0 :	ℓ/a more than 8.0 :
$\phi_t = \dfrac{w\ell ea}{JG}\left(\dfrac{\ell}{2a} - 1 + \dfrac{a}{\ell}\right)$	$\phi_t = \dfrac{w\ell ea}{JG}\left(\dfrac{\ell}{8a} - \dfrac{a}{\ell}\right)$	$\phi_t = \dfrac{w\ell ea}{JG}\left(\dfrac{\ell}{8a} - \dfrac{1}{2}\right)$

**Interpolation formulas for bending moment in each flange at F
and maximum total twist angle at location ϕ.***

Case 4. Cantilever. ℓ/a more than 0.5 and less than 3.0 :

$$M_f = \frac{w\ell ea}{h}\left[0.041 + 0.423\frac{\ell}{a} - 0.068\left(\frac{\ell}{a}\right)^2\right]$$

$$\phi_t = \frac{w\ell ea}{JG}\left[-0.023 + 0.029\frac{\ell}{a} + 0.086\left(\frac{\ell}{a}\right)^2\right]$$

Case 5. Simple span. ℓ/a more than 1.0 and less than 6.0 :

$$M_f = \frac{w\ell ea}{h}\left[0.097 + 0.094\frac{\ell}{2a} - 0.0255\left(\frac{\ell}{2a}\right)^2\right]$$

$$\phi_t = \frac{w\ell ea}{JG}\left[-0.032 + 0.062\frac{\ell}{2a} + 0.052\left(\frac{\ell}{2a}\right)^2\right]$$

Case 6. Continuous spans of equal length, equally loaded.
ℓ/a more than 2.0 and less than 8.0 :

$$M_f = \frac{w\ell ea}{h}\left[0.005 + 0.342\frac{\ell}{4a} - 0.078\left(\frac{\ell}{4a}\right)^2\right]$$

$$\phi_t = \frac{w\ell ea}{2JG}\left[-0.029 + 0.266\left(\frac{\ell}{4a}\right)^2\right]$$

* Note. See footnote to Table 8.1

8.1 and 8.2 for bending of the complete beam. Likewise, the maximum beam moments and flange bending moments occur at identical locations, and the direct stresses in the flanges due to the two causes are additive. Thus Eqs. (8.11) and (8.12), as applied to the short cantilever, may also be applied to any of the other five loading and support conditions.

The design of intermediate and long W beams in combined bending and torsion will be feasible only if the torsional moments are small. Otherwise, torsional deflections will be prohibitively large. Direct flange stresses should be checked for combined bending and torsion, and the maximum twist determined. Away from restrained ends the torsional shear stress, of the pattern shown in Fig. 8.4(a), may be calculated by Eq. (8.9), but it is bound to be small and of little design significance if the total twist angles are small.

Several design examples will now illustrate the foregoing procedures and point up situations where W shapes are satisfactory in combined bending torsion, as well as other cases where the reverse is true.

Example 8.1

A 3-kip pull is applied at any angle, tangential to the circumference of a 20-in. diameter, as shown, at the top of a 20-in.-long W beam that is fixed at its base. Select beam size using A36 steel, $F_y = 36$ ksi. Obviously, the design may be based on the pull being directed so as to cause bending about the weak axis of the beam, as shown at the bottom of the drawing below.

Refer to case 1, Table 8.1, and assume that l/a is less than 0.5—to be checked after selection.

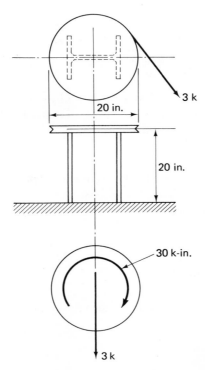

Try a W 12 beam and assume that $h = 11.5$ in.

$$M_f = \frac{Pel}{h} = \frac{3 \times 10 \times 20}{11.5} = 52.0 \text{ kip-in.}$$
$$M_y = Pl = 3 \times 20 = 60.0 \text{ kip-in.}$$

By Eq. (8.12), the required section modulus S_y is determined:

$$S_y = \frac{60.0 + 2 \times 52.0}{24} = 6.83 \text{ in.}^3$$

Try W 12 × 36:

$$S_y = 7.77 \text{ in.}^3$$
$$J = 0.83 \text{ in.}^4$$
$$C_w = 873 \text{ in.}^6$$
$$h = 12.24 - 0.54 = 11.70 \text{ in.}$$

Check L/a by Eq. (8.10):

$$a = 1.61 \sqrt{\frac{873}{0.83}} = 52.2 \text{ in.}$$
$$\frac{l}{a} = \frac{20}{52.2} = 0.38 < 0.50 \qquad \text{OK}$$

Check direct stress in flange:

$$f_b = \frac{60.0}{7.77} + \frac{2 \times 3 \times 10 \times 20}{11.7 \times 7.77} = 20.9 \text{ ksi} < 24 \qquad \text{OK}$$

Check the maximum shear stress:

$$A_f = 6.57 \times 0.54 = 3.55 \text{ in.}^2$$
$$V_{max} = \frac{3}{2} + \frac{3 \times 10}{11.7} = 4.06 \text{ kips}$$
$$f_v = \frac{1.5 \times 4.06}{3.55} = 1.72 \text{ ksi} < 14.5 \qquad \text{OK}$$

Note that in spite of the very short length of the beam and the orientation that makes the direct flange shear forces due to torsion and bending directly additive, the maximum shear stress is not significant.

Example 8.2

Redesign for Ex. 8.1, changing W section to a pipe section. Ignore torsion in preliminary selection and assume $F_b = 22$ ksi.

Section modulus required is

$$S = \tfrac{60}{22} = 2.73 \text{ in.}^3$$

(Refer to AISCM, page 1-102.)

Try 4-in. standard pipe:

$$S = 3.21 \text{ in.}^3 \qquad \text{O.D.} = 4.5 \text{ in.}$$
$$A = 3.17 \text{ in.}^2 \qquad \text{Mean diameter: } d_m = 4.5 - 0.237 = 4.26 \text{ in.}$$
$$t = 0.237 \text{ in.}$$

Check shear stress due to torsion and bending. (The shear shape factor for a circular tube in bending is 1.33).

Due to bending,

$$f_v = \frac{1.33 \times 3}{3.17} = 1.26 \text{ ksi}$$

Due to torsion,

$$f_v = \frac{3 \times 10}{2 \times 0.237 \times \pi \times 2.13^2} = 4.44 \text{ ksi}$$

Due to both bending and torsion,

$$f_v = 5.7 \text{ ksi} < 14.5 \qquad \text{OK}$$

Direct stress due to bending:

$$f_b = \frac{60}{3.21} = 18.7 \text{ ksi}$$

Reduced allowable direct stress because of shear stress [Eq. (8.13)]:†

$$F_b = F_{ra} = \left[1 - \left(\frac{4.44}{22}\right)^2\right]22 = 21.1 \text{ ksi} > 18.7 \qquad \text{OK}$$

Examples 8.1 and 8.2 showed that either a W or pipe section could serve satisfactorily in combined bending and torsion for a short cantilever member. The required W section weighed three times more than the pipe. In the design of relatively long members in combined bending and torsion, such as might be required for a highway direction sign attached to a single vertical member, the pipe is the only suitable member because of the unduly large torsional deflections that would result from use of a W section.

Examples 8.3 and 8.4 illustrate the combined bending and torsion problem in the design of a continuous spandrel beam supporting an exterior wall. It is assumed that the framing situation does not permit reduction or elimination of the torsion component by means of a laterally contiguous floor slab or framed beams.

Example 8.3

Continuous 24-ft spans of a spandrel beam carry a wall load of 600 lb/ft, 4 in. from the center of the beam. Use A36 steel and a W beam. Wall support brackets will not be considered as adding to the beam section.

---

†*Note regarding combined shear and direct stress:* The AISCS does not provide any limitation on direct stress combined with shear, except in the case of connection design. To keep the maximum principal stress less than F_a, the allowable direct stress, the solution of the quadratic principal stress formula may be avoided for the simple case of a single direct stress component by the determination of a reduced allowable direct stress (F_{ra}) as follows:

$$F_{ra} = \left[1 - \left(\frac{f_v}{F_a}\right)^2\right]F_a \tag{8.13}$$

If f_v/F_a is less than 0.2, the effect of shear stress on the allowable stress may be ignored. The case in Ex. 8.2 is borderline.

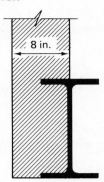

A preliminary trial selection will be chosen by designing for bending alone, but at a greatly reduced allowable stress, say, one third of 22 or 7.33 ksi. Because of lack of lateral support, the provisions of AISCS, Sec. 1.5.1.4.1, for reduced end moment in the continuous beam cannot be applied. Design moment, assuming a beam weight of 30 lb/ft, is calculated:

$$M_x = \frac{0.63 \times 24^2 \times 12}{12} = 362.9 \text{ kip-in.}$$

Required section modulus for trial beam 7.33 ksi,

$$S_x = \frac{362.9}{7.33} = 49.5 \text{ in.}^3$$

Refer to AISCM, page 2-10, and choose a W 14 × 38 as a trial, for which

$$S_x = 54.7$$
$$S_y = 7.86$$
$$J = 0.796$$
$$C_w = 1230.0$$
$$h = 14.12 - 0.51 = 13.61 \text{ in.}$$

By Eq. (8.10),

$$a = 1.61 \sqrt{\frac{1230}{0.796}} = 63.3 \text{ in.}$$

$$\frac{l}{a} = \frac{288}{63.3} = 4.55$$

(More than 2.0, less than 8.0; hence use interpolation formulas of Table 8.2, case 6.) Using the interpolation formula, the maximum moment in a flange due to torsional bending is found:

$$M_f = \frac{(0.64)(288)(4.0)(63.3)}{(12)(13.61)} \left[0.005 + 0.342 \left(\frac{4.55}{4} \right) - 0.078 \left(\frac{4.55}{4} \right)^2 \right] = 83.8 \text{ kip-in.}$$

The stress due to torsional bending alone is

$$f_{bt} = \frac{83.8 \times 2}{7.86} = 21.3 \text{ ksi}$$

too great, obviously, for the torsion bending stress alone. To make a better selection, assume $M_f = 84$ kip-in., as in this case, $S_x/S_y = 54.7/7.86 = 6.96$. Then, by Eq. (8.11),

$$\text{reqd } S_x = \frac{362.9 + 2 \times 6.96 \times 84}{22} = 69.9$$

requiring, by AISCM, page 2-10, a W 21 × 44. But, noting that S_x/S_y would be more than 10, the requirement for S_x would be escalated. It is obviously desirable to stay with a wider and less deep cross section. Noting, also, that the next group of sections heavier than the W 14 × 30 trial selection have $S_x/S_y = 5.5$ for the median of the group (page 1-36, AISCM), the required S_x would be reduced to

$$\frac{362.9 + 2 \times 5.5 \times 84}{22} = 58.5 \text{ in.}^3$$

Try a W 14 × 43, for which $S_x = 62.7$.
    Other needed properties of the W 14 × 43 are

$$S_y = 11.3$$
$$J = 1.05$$
$$C_w = 1950.0$$
$$h = 13.68 - 0.53 = 13.15$$
$$a = 1.61\sqrt{\frac{1950}{1.05}} = 69.4$$
$$\frac{l}{a} = \frac{288}{69.4} = 4.15$$

Again, use interpolation formulas from Table 8.2, case 6, for flange moment:

$$M_f = \frac{(0.64)(288)(4)(69.4)}{(12)(13.15)}\left[0.005 + 0.342\left(\frac{4.15}{4}\right) - 0.078\left(\frac{4.15}{4}\right)^2\right] = 89.4 \text{ kip-in.}$$

Stress due to bending moment M_x:

$$f_{bx} = \frac{0.64 \times 24^2 \times 12}{12 \times 62.7} = 5.9 \text{ ksi}$$

Stress due to torsional flange bending:

$$f_{bt} = \frac{89.4 \times 2}{11.3} = 15.8 \text{ ksi}$$

Total direct stress due to combined bending and torsion:

$$f_b = 5.9 + 15.8 = 21.7 \text{ ksi} < 22 \qquad \text{OK}$$

provided that 5.9 ksi is less than permissible stress for the laterally unsupported beam. Assume $C_b = 1$ (AISCS, Sec. 1.5.1.4.6):

$$\frac{d}{A_f} = 3.24 \qquad \text{(AISCM, page 1-37)}$$

F_b is no less than $12{,}000/(288 \times 3.24) = 12.86 > 5.9$,   OK.
    Check maximum twist at center by interpolation formula from Table 8.2, case 6:

$$\phi_t = \frac{(0.64)(288)(4)(69.4)}{(12)(2)(1.05)(11{,}200)}\left[-0.029 + 0.266\left(\frac{4.15}{4}\right)^2\right] = 0.047 \text{ radian}$$

Thus, if the masonry wall lacked internal restraint (if placed to a height of 50 in. before setting up of mortar), a point 50 in. above the top of the beam (assuming the beam to be about half loaded) would tend to deflect outward:

$$\tfrac{1}{2} \times 0.057 \times 50 = 1.16 \text{ in.}$$

Since the deflection would increase gradually as the wall was placed, it might be partially offset by progressive correction as bricks or blocks were placed. Thus, for eccentricities of *only an inch or two*, the use of W sections for spandrel beams, designed as above for combined bending and torsion, would be feasible without excessive twist.

Example 8.4

Alternative design using a box (rectangular tube) section for the same support and load conditions as Ex. 8.3. See sketch.

Initial design will be for full bending moment at $F_b = 22$ ksi, neglecting torsion. Assume weight of member at 0.02 kip/ft:

$$M_x = \frac{0.62 \times 24^2 \times 12}{12} = 375.1 \text{ kip-in.}$$

Required section modulus:

$$S_x = \frac{357.1}{22} = 16.23 \text{ in.}^3$$

Refer to the AISCM, page 1-104, for properties of rectangular structural tubing. Try TS $12 \times 4 \times 0.250$:

$$S_x = 20.5 \text{ in.}^3, \qquad f_b = \frac{357.1}{20.5} = 17.4 \text{ ksi}$$

$$t = 0.25 \text{ in.}$$
$$h = 12.0 - 0.25 = 11.75 \text{ in.}$$
$$b = 4.0 - 0.25 = 3.75 \text{ in.}$$
$$A_o = 3.75 \times 11.75 = 44.1 \text{ in.}^2$$

Check shear stress due to combined bending and torison. Due to beam bending,

$$f_{vb} = \frac{0.61 \times 12}{0.5 \times 12} = 1.24 \text{ ksi}$$

Due to torsion,

$$M_t = \frac{wel}{2} = \frac{0.62 \times 4 \times 288}{12 \times 2} = 29.8 \text{ kip-in.}$$

$$f_{vt} = \frac{29.8}{2 \times 44.1 \times 0.25} = 1.35 \text{ ksi}$$

$$f_v = 1.24 + 1.35 = 2.6 \text{ ksi} < 14.5 \qquad \text{OK}$$

Check for lateral buckling. AISCS, Sec. 1.5.1.4.4, allows $F_b = 0.6F_y$ if l/b is less than $2500/F_y$, which is 69.4 for $F_y = 36$ ksi (AISCM, page 5-68) (b is taken as out to out); $\frac{288}{4} = 72$, which slightly exceeds the allowable, but since the stress due to bending is 17.4, appreciably less than 22, the design is satisfactory for lateral buckling. Calculate the maximum twist angle at the center of span. The torsion constant, by Eq. (8.7), is

$$J = \frac{4 \times 44.1^2}{2(11.75 + 37.5)/0.25} = 62.7 \text{ in.}^4$$

Contrast this with $J = 1.05$ for the W 14×43 beam selection in Ex. 8.3. The *average* torsional moment between one end and the center of the span is $wel/4$ and

$$\phi_t = (\theta_{\text{avg}})\left(\frac{l}{2}\right) = \frac{wel^2}{8JG}$$

$$\phi_t = \frac{0.62 \times 4 \times 288^2}{12 \times 8 \times 62.7 \times 11{,}200} = 0.003 \text{ radian}$$

to be contrasted with ϕ_t of 0.057 radian for the W beam selection of Ex. 8.3.

The foregoing examples have demonstrated design procedures for both open and closed sections and have shown the strength and stiffness advantages of the closed box section in resisting torsion loads.

8.4   BIAXIAL BENDING AND LATERAL-TORSIONAL BUCKLING

Properties for structural sections as listed in the AISCM are in terms of the xx and yy axes and, except for the angle and zee† sections, the xx and yy axes are also the *principal* axes of the cross sections, as will always be the case if one of the two axes is an axis of symmetry. As explained in Chapter 3, if lateral support is provided, either continuously or at the locations of applied concentrated load, a beam of any shape may be designed on the basis of simple bending theory. If lateral support is not provided, the possibility of lateral-torsional buckling about the weakest principal axis is always present, and the design procedure to be recommended here requires the determination of the orientation of the principal axes, in cases where they are not already known, and the calculation of the principal moments of inertia, designated herein as I_1 and I_2. After these are determined, and the effect of lateral-torsional

---

†No longer listed in the AISCM.

buckling in reducing the allowable stress in bending about the strong bending axis is estimated, the procedure parallels that of Chapter 3.

In the general case, to determine the principal moments of inertia, it will be necessary to calculate the product of inertia, I_{xy}. The reader should refresh his acquaintance with this parameter by reference to his text on strength of materials. It will be recalled that

$$I_x = \int y^2 \, dA, \qquad I_y = \int x^2 \, dA, \qquad I_{xy} = \int xy \, dA$$

Note that I_x and I_y are always positive quantities, but that I_{xy} may be either positive or negative, and that for the same section this will depend on the arbitrary way in which the positive directions of x and y chosen. If, as is so common, the structural shape is made of rectangular component parts, the contribution of any one rectangle to I_{xy} may be determined by the *parallel-axis theorem*,

$$I_{xy} = I_{xyo} + Ax_oy_o \tag{8.14}$$

If the x and y axes include one that is an axis of symmetry, they are then principal axes, and I_{xyo} is zero. Thus, for a rectangular component, I_{xyo} is different from zero only if the sides are tilted at an angle to the x and y axes, as illustrated in Fig. 8.6, which is explanatory of Eq. (8.15):

$$I_{xyo} = \frac{b^3t - bt^3}{12} \sin \theta \cos \theta \tag{8.15}$$

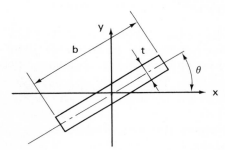

Fig. 8.6

If b is relatively large in relation to t, the term bt^3 may be omitted from Eq. (8.15) with but little error. It should be noted that when θ is either zero or 90°, I_{xyo} is zero. Another useful relationship is the fact that, regardless of the orientation of x and y, the sum of I_x and I_y is a constant. Thus, also,

$$I_x + I_y = I_1 + I_2 \tag{8.16}$$

Equation (8.16) is useful in the determination of the principal moments of inertia of an angle, using the information listed in the AISCM, as illustrated by Ex. 8.5:

Example 8.5

Referring to page 1-59 of the AISCM, determine the principal moments of inertia of an L 6 × 4 × ½.

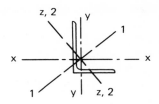

$$I_x = 17.4 \text{ in.}^4, \qquad I_y = 6.27 \text{ in.}^4$$

and the minimum radius of gyration about the principal axis zz is listed as 0.870 in. The area A is given as 4.75 in.^{2}, and from the relationship $I = Ar^2$, the minimum moment of inertia, I_2, is

$$I_2 = 4.75 \times 0.87^2 = 3.60 \text{ in.}^4$$

Equation (8.16) now provides the calculation of I_1:

$$I_1 = 17.4 + 6.27 - 3.6 = 20.1 \text{ in.}^4$$

The complete problem of calculating section properties in the more general case for which no handbook information is available will be illustrated in Ex. 8.6. If the orientation of the principal axes is not known, it may be determined by Eq. (8.17), in which θ is the angle between the x axis and the principal axes. Only one of the angles need be calculated since they are 90° apart:

$$\tan 2\theta = \frac{2I_{xy}}{I_y - I_x} \tag{8.17}$$

The magnitudes of the principal moments of inertia are given by

$$I_1, I_2 = \frac{I_x + I_y}{2} \pm \sqrt{\left(\frac{I_x - I_y}{2}\right)^2 + I_{xy}^2} \tag{8.18}$$

Stresses may now be determined by resolving the loads into components in the principal planes and superposing the stresses as calculated by the ordinary beam stress formula applied successively to the bending moments about each of the two principal axes, as in the case treated in Chapter 3 [Eq. (3.12)], which was applicable when the handbook x and y axes were also principal axes. A scale layout will be helpful to determine distances to extreme fibers in the stress calculations.

However, if there is no lateral support, as is always the case if applied loads cause stress about both principal axes, there will be a reduction in allowable stress required for bending about the strong axis because of the lateral-torsional buckling effect. The AISCS interaction procedure, also explained and illustrated in Chapter 3 [Eq. (3.13)], may then be applied, provided that the reduced allowable stress for bending about the strong axis is determined, as will be discussed later.

Example 8.6

Determine the section properties about the x and y axes, and about the principal axes, for the shape shown. As a close approximation in the calculations it is assumed that the section is made up of two rectangular parts with breadths of 8 and 10 in., respectively.

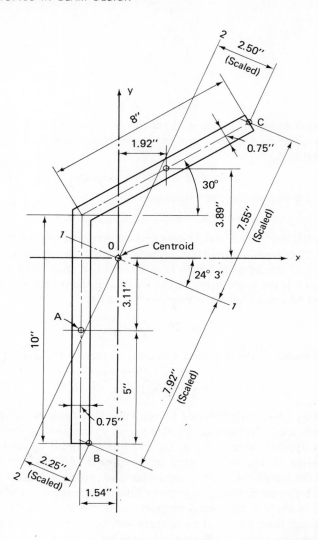

Area:

$$A = 0.75(8.0 + 10.0) = 13.5 \text{ in.}^2$$

Locate the neutral axes. Use point A, the centroid of the 10-in. segment, as a reference origin:

$$\bar{x}_A = \frac{6.0 \times 3.46}{13.5} = 1.54 \text{ in.}$$

$$\bar{y}_A = \frac{6.0 \times 7.0}{13.5} = 3.11 \text{ in.}$$

Having located the centroid, this becomes the origin of x and y distances to be used in calculation of the moments of inertia, I_x, I_y, and I_{xy}.

Determine I_x:

$$6.0 \times 3.89^2 + \frac{6.0 \times 4.0^2}{12} = 98.79$$

$$7.5 \times 3.11^2 + \frac{7.5 \times 10.0^2}{12} = \underline{135.06}$$

$$I_x = 233.85 \text{ in.}^4$$

Determine I_y:

$$6.0 \times 1.92^2 + \frac{6.0 \times 6.92^2}{12} = 46.06$$

$$7.5 \times 1.54^2 + \frac{7.5 \times 0.75^2\dagger}{12} = \underline{18.14}$$

$$I_y = 64.20 \text{ in.}^4$$

Determine I_{xy}:

To calculate I_{xyo} of the 8 × 0.75 in. segment, by Eq. (8.15), for $\theta = 30°$,

$$\begin{aligned} \sin \theta &= 0.500 \\ \cos \theta &= 0.866 \end{aligned} \quad \text{(AISCM, pages 6-32, 6-33)}$$

Equation (8.15), for calculation purposes, may be more conveniently expressed as

$$I_{xyo} = \frac{bt(b^2 - t^2)}{12} \sin \theta \cos \theta$$

The t^2 term will be included for sake of completeness:

$$I_{xyo} = \frac{6.0(8.0^2 - 0.75^2)}{12}(0.5 \times 0.866) = 13.73 \text{ in.}^4$$

Thus for the complete section, by Eq. (8.14),

$$I_{xy} = (7.5)(-3.11)(-1.54) + (6.0)(+3.89)(+1.92) + 13.73 = +94.47 \text{ in.}^4$$

Orientation of the principal axes is obtained by use of Eq. (8.17),

$$\tan 2\theta = \frac{2(+94.47)}{64.20 - 233.85} = -1.114$$

From AISCM, page 6-35, $2\theta = -48°5'$; hence $\theta = -24°3'$, and

$$\begin{aligned} \sin \theta &= -0.4075 \\ \cos \theta &= 0.9132 \end{aligned}$$

The principal moments of inertia are now calculated by Eq. (8.18):

$$I_1, I_2 = \frac{233.85 + 64.20}{2} \pm \sqrt{\left(\frac{233.85 - 64.20}{2}\right)^2 + 94.47^2}$$

$$I_1 = 275.99 \text{ in.}^4$$
$$I_2 = 22.07 \text{ in.}^4$$

Suppose, now, that lateral support is provided for the section of Ex. 8.6, either continuously, if the load is continuous, or by rods attached at all load locations, as

†This term could have been omitted with but little error.

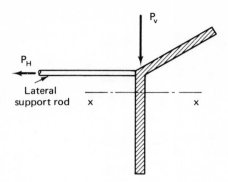

Fig. 8.7

shown in Fig. 8.7. With such support, bending is forced to be about the xx axis and the stress may be calculated by the usual Mc/I formula. The load-carrying capacity is increased by such lateral support, and, if the use of supports is optional, the cost of supports can be weighed against the cost of a heavier member that would be required if such supports were lacking. The force P_H in the support may be calculated by

$$P_H = \frac{P_V I_{xy}}{I_x} \tag{8.19}$$

The design of a laterally unsupported unsymmetrical section is illustrated next.

Example 8.7

A member having the properties and cross section of Ex. 8.6 has a span of 18 ft and is loaded vertically at the third points. If the allowable stress is 22 ksi (A36 steel), what load can the member safely carry if lateral supports are provided at the load locations as shown in the sketch?

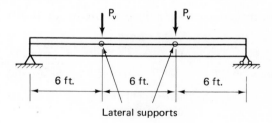

Lateral supports

Calculate bending moment:

$$M_x = 72 P_V \text{ kip-in.}$$

Referring to Ex. 8.6, the maximum stress in tension, 8.11 in. below the neutral axis, will determine the load capacity:

$$f_b = 22 = \frac{72 P_V \times 8.11}{233.85}$$

Solving,

$$P_V = 8.81 \text{ kips}$$

In trying to bend laterally, about its weak axis, the member would obviously tend to deflect to the right. Thus the stress in the lateral support rods would be tension in the amount

$$P_H = \frac{8.81 \times 94.47}{233.85} = 3.56 \text{ kips}$$

If there is no lateral support, the allowable stress in bending about the strong axis will probably need to be reduced to provide safety against lateral-torsional buckling. The AISCS covers only specific cross sections. In the present case a conservative estimate of the critical moment that will cause lateral-torsional buckling is provided by

$$M_{cr} = \frac{\pi}{l}\sqrt{JGEI_2} \tag{8.20}$$

For steel, $\sqrt{GE} = 18,000$, and

$$M_{cr} = \frac{18,000\pi}{l}\sqrt{JI_2}$$

Although overconservative for bent beams, it is convenient in situations not covered by the specifications to convert the beam-buckling problem into an equivalent column problem and thus permit direct use of tables of column allowable stresses. The buckling stress in compression is calculated:

$$f_{cr} = \frac{18,000\pi c_c}{lI_1}\sqrt{JI_2} \tag{8.21}$$

where c_c is the distance from the 1-1 principal axis to the extreme fiber in compression.

Table 2, Appendix A, on page 5-94 of the AISCS, gives column buckling stresses, assuming elastic behavior, but divided by the long column factor of safety of $\frac{23}{12}$. Thus, if the beam buckling stress by Eq. (8.21) is divided by $\frac{23}{12}$, one can enter Table 2 with the corresponding stress and read out the "equivalent" column slenderness ratio, Kl/r. This can be used to determine a safe beam buckling stress by use of the allowable stress tables for columns, as given in the AISCM. The result is overconservative for two reasons: (1) Eqs. (8.20) and (8.21) neglect the bending contribution to torsional resistance, and (2) the factor of safety used in the column tables is greater than that specified for beams. The equivalent column procedure is now illustrated in Ex. 8.8.

Example 8.8

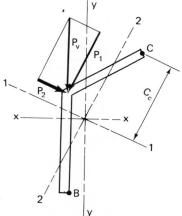

Same as Ex. 8.7 but without any lateral support. Resolve P_V into principal plane components P_1 and P_2:

$$P_1 = P_V \cos \theta = 0.9132 \, P_V$$
$$P_2 = P_V \sin \theta = 0.4075 \, P_V$$

Stress at $B = 22$ ksi (trial):

$$22 = \frac{0.9132 P_V \times 6 \times 12 \times 7.92}{275.99} + \frac{0.4075 P_V \times 6 \times 12 \times 2.25}{22.07}$$
$$22 = 1.887 P_V + 2.991 P_V = 4.878 P_V$$
$$P_V = 4.51 \text{ kips}$$

Check selection by AISCS interaction formula procedure (refer to Chapter 3, p. 56). Referring to Ex. 8.6, the scaled distance from the 1-1 principal axis to the location of average maximum compression stress (C), distance c_c, is 7.55 in. The torsion constant, J, by Eq. (8.8), is

$$J = \frac{(10 + 8) \, 0.75^3}{3} = 2.53 \text{ in.}^4$$

By Eq.(8.21),

$$f_{cr} = \frac{18,000 \times 3.1416 \times 7.55}{216 \times 275.99} \sqrt{2.53 \times 22.07} = 53.5 \text{ ksi}$$

The equivalent column buckling stress, divided by $\frac{23}{12}$, is

$$F'_e(\text{equiv}) = \frac{53.5 \times 12}{23} = 27.91 \text{ ksi}$$

Referring to Table 2, AISCS, page 5-94, the equivalent

$$\frac{Kl}{r} = 73.1$$

(*Note:* For remarks on load and support locations in Exs. 8.7 and 8.8, refer to Sec. 8.5.)
Now enter the Allowable Column Stress Table 1-36, page 5-84, AISCS, and obtain

$$F_a = 16.10 \text{ ksi}$$

for a column with $Kl/r = 73.1$. This will be a safe allowable maximum stress for the beam buckling condition of this problem. It is obvious that the value of $P_V = 4.51$, based on a maximum stress of 22 ksi for bending about both axes, is too great. Try $P_V = 4$ kips and calculate bending stresses f_{b1} and f_{b2} separately:

$$f_{b1} = \frac{0.9132 \times 4 \times 72 \times 7.55}{275.99} = 7.19 \text{ ksi}$$
$$f_{b2} = \frac{0.4705 \times 4 \times 72 \times 2.50}{22.07} = 13.29 \text{ ksi}$$

Applying the interaction formula,

$$\frac{7.19}{16.10} + \frac{13.29}{22} = 1.05 > 1 \qquad \text{NG}$$

To avoid repeating the foregoing calculations, the load that will bring the right side of the interaction formula down to 1 can be approximated as equal to $4/1.05 = 3.81$ kips.

The approach used in Ex. 8.8 may also be applied to the design of box girders of such slender proportions as to fall outside the scope of the simplified rules suggested in Chapter 3, or to solid rectangular bars, or to other closed sections that fall outside specification rules. Example 8.9 will illustrate.

Example 8.9

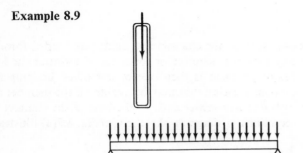

Determine the capacity under uniform load of a TS $12 \times 2 \times 0.250$ in. rectangular tube used as a laterally unsupported simple beam with a span of 20 ft.

$$\frac{l}{b} = \frac{20 \times 12}{2} = 120$$

Therefore stress in bending must be reduced, as the AISCS permits $F_b = 22$ ksi for l/b only up to 69.4 (AISCS, Sec. 1.5.1.4.4, page 5-68). From AISCM, page 1-104, the properties of this box tube are

$$I_x = 88.3 \text{ in.}^4$$
$$I_y = 4.51 \text{ in.}^4$$
$$S_x = 14.7 \text{ in.}^3$$

The torsion constant is calculated by Eq. (8.7):

$$J = \frac{2 \times 1.75^2 \times 11.75^2}{(11.75 + 1.75)/0.25} = 15.66 \text{ in.}^4$$

Calculate critical stress by Eq. (8.21):

$$f_{cr} = \frac{18,000\pi}{240 \times 14.7}\sqrt{15.66 \times 4.51} = 134.7 \text{ ksi}$$

The equivalent column buckling stress, divided by $\frac{23}{12}$, is

$$F'_e = \frac{134.7 \times 12}{23} = 70.3 \text{ ksi}$$

and from Table 2, AISCS, page 5-94, the equivalent

$$\frac{Kl}{r} = 46.1$$

Allowable Column Stress Table 1-36, page 5-84, AISCS, gives an allowable column stress $F_a = 18.69$ ksi, and this may also be used as a conservative estimate for the allowable maximum stress in bending in this example.

8.5 SHEAR CENTER

If certain structural sections, such as the channel and angle, are loaded through their centroidal axis without any torsional support or torsional restraint at the load points, they will twist. The design problem is then one of combined bending and torsion, as covered in Sec. 8.3, a complication that may be avoided if the member can be loaded and supported through its *shear-center axis*. In the case of the channel the three-dimensional free-body equilibrium diagram sketched in Fig. 8.8(a) illustrates

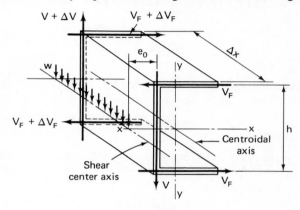

Fig. 8.8(a)

a short segment, Δx in length, cut from a beam shown loaded through the shear-center axis, so as to avoid twist. Since the loads are parallel to a principal axis of the cross section, the member will be in simple bending and without twist. The torsional couple $\Delta V_F h$ is held in equilibrium by the opposed torsional couple $w(\Delta x)e_o$, as illustrated. Referring to Fig. 8.8(b), the distance from the middle plane of the web to

Fig. 8.8(b)   Shear center of channel.

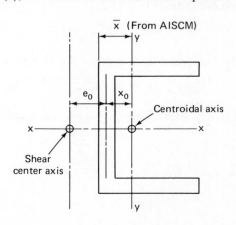

the shear-center axis is

$$e_o = \frac{x_o h^2}{4r_x^2} \tag{8.22}$$

where r_x is the radius of gyration about the xx axis. Equation (8.22) applies to channels with nonparallel flange faces as well as parallel.

If a channel supports beams that frame into it, the arrangement in Fig. 8.9(a) is preferred over that of Fig. 8.9(b). The channel may be assumed to be without twist and loaded through its shear-center axis, in either case, provided the supported beams are designed for span l measured in each case from the shear-center axis. In addition, in Fig. 8.9(a) connecting bolts should be located as near as possible to a vertical line

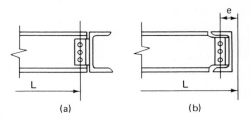

Fig. 8.9 Alternative framing arrangements for channel beam.

(a)   (b)

passing through the shear-center axis, whereas in Fig. 8.9(b) the connection should preferably be designed for an eccentricity of load equal to the distance from the shear-center axis to the bolt line. The channel as a spandrel beam loaded through the shear center is treated in Ex. 8.10.

Example 8.10

Select a channel as a 22-ft simple span spandrel beam for a load of 2 kips/ft, including weight of channel, and locate 8-in. wall so as to eliminate torsion of beam.

Maximum bending moment:

$$M_x = \frac{2 \times 22^2 \times 12}{8} = 1425 \text{ kip-in.}$$

Required section modulus, assuming $F_b = 22$ ksi:

$$S_x = \frac{1425}{22} = 66 \text{ in.}^3$$

Referring to beam selection tables, AISCM, page 2-9, MC 18 × 51.9 with $S_x = 69.7$ is OK, as is the lighter weight W 21 × 44, but if it is desired to hide the beam by the concrete wall, the channel offers advantages in the elimination of the combined bending and torsion problem, taking advantage of the shear-center location, as shown in the sketch. If laterally unsupported, d/A_f (AISCM, page 1-52) is 7.02, and the allowable stress is

$$F_b = \frac{12{,}000}{264 \times 7.02} = 6.48 \text{ ksi}$$

Thus, during construction, at least temporary lateral support would be needed. (Note that the beam selection tables list the maximum span, L_u, for which no lateral support is needed, in this case, 6.6 ft.) From AISCM, page 1-52, the properties of the MC 18 × 51.9 are

$$S_x = 69.7, \qquad r_x = 6.41, \qquad \bar{x} = 0.858$$

[In Eq. (8.22), $x_o = \bar{x} - (t_w/2) = 0.858 - 0.300 = 0.558$ in.]

$$h = d - t_f = 18.0 - 0.625 = 17.375 \text{ in.}$$

By Eq. (8.22),

$$e_o = \frac{0.558 \times 17.375^2}{4 \times 6.41^2} = 1.02 \text{ in.}$$

suggesting the relative channel and wall location as shown in the sketch.

Another situation conducive to use of a channel beam may occur if loads must be suspended by hanger rods, as shown in Fig. 8.10. These may be located at the shear center and again the combined bending and torsion problem is avoided.

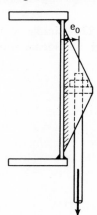

Fig. 8.10 Channel girder supporting hanger loads.

Formulas for the shear-center location of other shapes and procedures for determining the location for shapes having no axis of symmetry are given in Reference 1.7. When a section consists of only two rectangular component parts, the shear center is at the intersection of their two middle planes, as shown for the angle and tee section in Fig. 8.11. Referring back to Exs. 8.7. and 8.8, it should be noted that loads

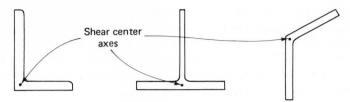

Fig. 8.11   Shear center locations for two-part sections.

and supports were both introduced at the shear-center axis, thus eliminating torsion from the problem.

8.6 COMPOSITE DESIGN

In composite design two different materials are joined together so as to utilize the properties of each to the best advantage. Thus when a reinforced concrete floor slab is attached to a supporting system of steel beams by means of shear connectors, as shown in Fig. 8.12, the entire system acts as a unit. The slab not only serves its usual purpose of supporting floor loads and transmitting these to the beams, but also provides a compression flange augmenting the top flange of the steel beams. The portion of the flange supplied by the concrete is either all or mostly in compression, utilizing concrete to its best advantage.

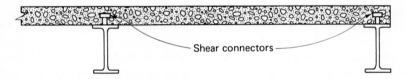

Fig. 8.12   Composite concrete slab and steel beam construction.

The AISCM, pages 2-135 to 2-195, provides general notes, design examples, and composite beam property tables for a wide variety of steel beams, both with and without cover plates, and with slab thicknesses varying from 4 to $5\frac{1}{2}$ in. Quoting from the manual, "Composite construction is appropriate for any loading. It is most efficient with heavy loading, relatively long spans, and beams spaced as far apart as permissible." Composite design is widely used and offers advantages of economy and of overall structural integrity. Because of the excellent coverage of the subject in the AISCM, further treatment in detail will herein be omitted.

8.7 CONTINUOUS BEAMS

The design of continuous beams will be considered briefly by two alternative procedures: (1) Allowable-stress design, and (2) plastic design, both permissible by the AISCS.

In the traditional allowable-stress procedure, an elastic continuous beam analysis is made to determine maximum positive and negative bending moments for various critical positions of live load. If the chosen W beam section meets the requirements of AISCS, Sec. 1.5.1.4.1, the beam need be "proportioned for $\frac{9}{10}$ of the negative moments produced by gravity loading which are maximum at points of support, provided that, for such members, the maximum positive moment shall be increased by $\frac{1}{10}$ of the average negative moments." This adjustment in moments is justified by the fact that when yielding starts at the supports, due to negative moment, the positive moments increase at a more rapid rate, and by the time the failure load is reached the positive and negative moments are more or less equalized.

In plastic design direct advantage is taken of the equalization of moments in the inelastic range, as covered by Part 2 of the AISCS. Referring to Fig. 3.7 and the accompanying discussion of inelastic behavior, the W section chosen must have flange and web thicknesses and meet lateral support requirements so that the plastic moment M_p will not only be approached, but will be maintained until the positive moments also reach M_p. The member size is chosen so that the continuous beam or frame will have a maximum strength at failure that will support the maximum expected "working" loads multiplied by a load factor of 1.7 (AISCS, Sec. 2.1).

Allowable-stress design is widely applicable to all types of structures and provides safe design against buckling, repeated load, and, with modifications in allowable stress and design moment, can provide to a degree a consistent relationship between design load and structural strength.

Plastic design is limited to continuous beams and frames and with modifications provides safe design against buckling. It should not be used if the number of cycles of repeated load (Appendix B, AISCS) reaches a number requiring a limit on the maximum stress range. Plastic design ensures an optimum of structural integrity, or "toughness," against failure—a desirable feature in earthquake- or blast-resistant structures. Moreover, especially in the case of continuous beams, the structural analysis for plastic design is simplified because it is *statically determinate*. A complete and authoritative commentary on plastic design is provided in American Society of Civil Engineers Manual 41, "Plastic Design in Steel."

Illustrative examples for both allowable stress and plastic design will be given. In Ex. 8.11 advantage is taken of the availability of tables of moments for continuous beams of two, three, or four spans of equal length in the AISCM. The design procedure is not essentially different if a statically indeterminate continuous beam analysis is required as provided by any one of several available methods.

Example 8.11

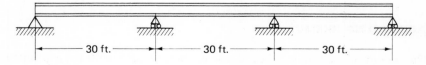

Select a W beam of A36 steel for a continuous beam of three equal spans of 30 ft each. A dead load of 1 kip/ft and live load of 2 kips/ft are assumed to be uniform. Use the allowable-stress procedure. Simple supports are assumed at extreme ends. Continuous lateral support will be provided by the floor system. Assume dead weight of the beam at 100 lb/ft. Calculate moments due to dead load (Case 36, AISCM, page 2-210). Positive moment in end span, $0.4l$ from end:

$$M = 0.080wl^2 = 0.080 \times 1.1 \times 30^2 \times 12 = 950 \text{ kip-in.}$$

Negative moment at supports:

$$M = -0.100wl^2 = -0.10 \times 1.1 \times 30^2 \times 12 = -1188 \text{ kip-in.}$$

Calculate moments due to live load (case 35, AISCM, page 2-210; only end spans loaded for maximum possible moment in end span).

Maximum possible moment $0.45l$ from end:

$$M = 0.1013wl^2 = 0.1013 \times 2.0 \times 30^2 \times 12 = 2188 \text{ kip-in.}$$

Negative moment for same load condition (not maximum):

$$M = -0.050wl^2 = -0.05 \times 2.0 \times 30^2 \times 12 = -1080 \text{ kip-in.}$$

Maximum negative moment (case 34, AISCM, page 2-210; one end span unloaded):

$$M = -0.1167wl^2 = -0.1167 \times 2.0 \times 30^2 \times 12 = -2521 \text{ kip-in.}$$

Make beam selection for $\frac{9}{10}$ maximum negative moment or for maximum positive moment increased by $\frac{1}{10}$ of the average negative moments (AISCS, Sec. 1.5.1.4.1). Negative design moment:

$$M_{\text{des}} = 0.9(1188 + 2521) = 3338 \text{ kip-in.}$$

Positive design moment (maximum dead and live load moments are conservatively assumed to be at the same location):

$$M_{\text{des}} = (950 + 2188) + 0.1\left(\frac{0 + 1080}{2}\right) = 3192 \text{ kip-in.}$$

Negative moment controls beam selection. Selection modulus required:

$$S_x = \frac{3338}{24} = 139 \text{ in.}^3$$

Referring to the allowable-stress beam selection tables, AISCM, page 2-8, a W 21 $\times$ 68 is selected. $S_x = 140$ in.^{3}. No limits are indicated on F_y' or F_y''; hence thickness ratios are satisfactory for the assumed allowable stress of $0.66F_y$ (AISCS, Sec. 1.5.1.4.1). Check shear stress. Maximum is at end of loaded middle span when one end span is unloaded (refer to cases 34 and 36, AISCM, page 2-210):

$$V_{\max} = 0.6 \times 1.1 \times 30 + 0.617 \times 2.0 \times 30 = 56.8 \text{ kips}$$

$$f_v = \frac{56.8}{21.13 \times 0.43} = 6.25 \text{ ksi} < 14.5 \qquad \text{OK}$$

(*Note:* If the shear stress is quite large, it might be desirable to check the direct stress in the web, adjacent to the flange fillet, at the interior support. The procedure used in Ex. 8.2 could be used, but this check is not required by AISCS and usually the stress will not be critical.)

Before considering the alternative plastic design procedure applied to the conditions of Ex. 8.11, the ultimate strength moment–load relationships for an end span and for an interior span of a continuous beam will be examined. In simple plastic-design analysis, it is assumed that the M–ϕ curve for the W section shown in Fig. 3.7 may be replaced by a two-straight-line approximation, as shown in Fig. 8.13.

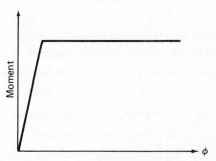

Fig. 8.13 Assumed beam behavior in plastic design analysis.

Relative angle between two nearby beam locations in maximum moment region

For the end span shown in Fig. 8.14, elastic analysis shows that M_p would first be reached at the support where the maximum moment occurs. According to Fig. 8.13, this moment would not change with increasing load, and the maximum load condition would be reached when a positive moment of M_p was also reached. It can be shown that the maximum positive moment is $0.414L$ from the simply supported end and that, for an end span,

$$M_p = 0.086wL^2 \tag{8.23}$$

Fig. 8.14 End span moments at ultimate plastic load.

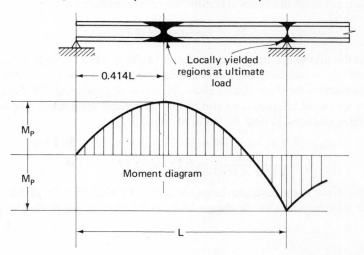

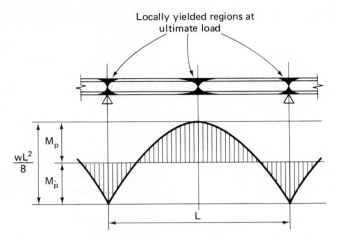

Fig. 8.15 Interior span moments at ultimate plastic load.

For an interior span, M_p would be reached simultaneously at both ends, and after additional increase in load would also be reached at the center. Since the total range of moment is the same as the center moment in a simple beam, it is readily seen (Fig. 8.15) that

$$M_p = \frac{wL^2}{16} \tag{8.24}$$

Example 8.12

Same loads and spans as for Ex. 8.11, but design by Part 2 of AISCS. Assume beam dead weight at 60 lb/ft. The total factored load on any one of the spans will be

$$1.7 \times 3.06 = 5.20 \text{ kips/ft}$$

Since the interior and end spans are of the same length, the beam selection will be determined by Eq. (8.23), which requires a greater M_p than Eq. (8.24). Required M_p, by Eq. (8.23), is

$$M_p = 0.086 \times 5.20 \times 30^2 \times 12 = 4831 \text{ kip-in.}$$

Required plastic modulus:

$$Z = \frac{4831}{36} = 134.2 \text{ in.}^3$$

Referring to Plastic Design Selection Tables, page 2-18, AISCM, try a W 24 × 55, with $Z = 134$ in.^{3}. The fractional percentage underrun can be tolerated. Check width–thickness ratios for compact section requirements of Sec. 2.7, AISCS:

$$\frac{b}{2t_f} = 6.96 < 8.5 \qquad \text{OK} \qquad \text{(AISCM, p. 1-31; AISCS, p. 5-60.)}$$

$$\frac{d}{t_w} = 59.5 < 68.7 \qquad \text{OK} \qquad \text{(AISCM, p. 1-31; AISCS, p. 5-80.)}$$

Check shear capacity. In the end span, adjacent to the interior support, maximum shear occurs equal to

$$V_{\max} = \frac{wL}{2} + \frac{M_p}{L} = (0.50 + 0.086)wL = 0.586wL$$
$$= 0.586 \times 5.2 \times 30 = 91.4 \text{ kips}$$

Shear capacity (AISCS, page 5-80) is

$$V = 19.8 \times 0.396 \times 23.55 = 184.6 > 91.4 \qquad \text{OK}$$

Example 8.13

Same as Ex. 8.12, but with supports relocated to reduce end spans to an optimum condition that would make the required M_p for the end spans the same as that for the center span.

Let x = length of end span. Then $90 - 2x$ = length of center span. By Eqs. (8.23) and (8.24),

$$0.086wx^2 = \tfrac{1}{16}w(90 - 2x)^2$$

This reduces to

$$x^2 - 137x + 3087 = 0$$

Solving the quadratic,

$$x = \tfrac{137}{2} \pm \tfrac{1}{2}\sqrt{137^2 - 4 \times 3087} = 68.5 \pm 40.1 = 28.4$$

Try two end spans of 28 ft–6 in. span, and a center span of 33 ft-0 in. Assume beam weight of 50 lb/ft. Factored load is

$$3.05 \times 1.7 = 5.19 \text{ kips/ft}$$

Required M_p for end span:

$$M_p = 0.086 \times 5.19 \times 28.5^2 \times 12 = 4346 \text{ kip-in.}$$

As a check, calculate M_p for center span:

$$M_p = \tfrac{1}{16} \times 5.19 \times 33^2 \times 12 = 4235 \text{ kip-in.}$$

(Slightly less, as exact optimum was not chosen.) Required plastic modulus:

$$Z = \tfrac{4346}{36} = 120.7 \text{ in.}^3$$

A beam size W 21×55, with $Z = 126$, is the least-weight selection (AISCM, page 2-18). Although the required Z has been reduced from 134.2 to 120.7, no weight saving results, but 3 in. has been taken off the building height.

Because of the discontinuous nature of available beam sizes and their corresponding plastic moduli, no general conclusions should be drawn from a comparison of Exs. 8.11, 8.12, and 8.13. Usually the beam selection by the Ex. 8.11 procedure would be more nearly the same as that by Ex. 8.12. Usually, the span spacing used in Ex. 8.13 would provide a weight saving in comparison with that of Ex. 8.12.

PROBLEMS†

8.1. Compare (a) torsional stiffness as measured by the torsion constant J for the following sections, which have approximately the same cross-sectional area, and (b) the torsional moment, for which the maximum shear stress is equal to 15 ksi. In the case of the W section, calculate the torsion constant by Eq. (8.8) and compare with the handbook value.

(a) Pipe 10 Std. ($A = 11.9$ in.²)
(b) TS 10 × 10 × 0.3125 ($A = 11.7$ in.²)
(c) W 10 × 39 ($A = 11.5$ in.²)

8.2. A vertical pipe supports a sign, as shown, which is to be designed for a wind force of 40 lb/ft². Neglecting weight of pipe and sign, and neglecting wind force on pipe, select a size for bending moment and check maximum shear stress due to combined torsion and bending. Redesign if necessary.

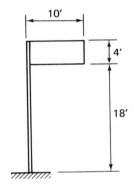

8.3. For the box section as shown in cross section, what is the torsional moment capacity for a maximum shear stress of 17 ksi? At this stress, what would the total angle of twist for a member 20 ft long amount to?

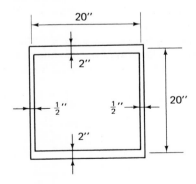

†A36 steel unless otherwise specified.

8.4. Using procedures applicable only to very short W sections, as described in connection with Fig. 8.5, determine the maximum normal and shear stress in the flange of the beam made up of three plates as shown.

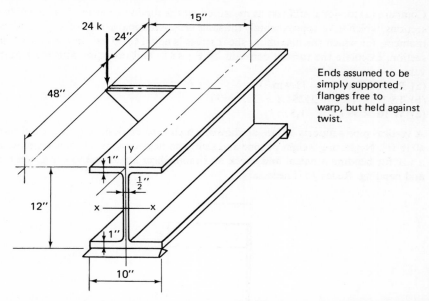

Ends assumed to be simply supported, flanges free to warp, but held against twist.

8.5. Similar to Example 8.3. Move the W section 1 inch laterally to reduce the eccentricity of load from 4 to 3 inches. Design as a simple beam instead of continuous and reduce the span from 24 to 20 ft.

8.6. Redesign the situation in Problem 8.5 using a standard rectangular structural tube. Procedure is similar to that of Ex. 8.4.

8.7. Determine the principal moments of inertia of an L $9 \times 4 \times \frac{3}{4}$ by the procedure used in Ex. 8.5.

8.8. A zee bar, cross section as shown, is used as a simple beam, 18 ft between supports, and is loaded at the third points with loads of 1200 lb each. To prevent lateral movement and force bending to remain in the plane of the web, lateral ties are introduced at the third points as shown. Determine the maximum stress due to bending and the force in the lateral ties. In which direction is the tie force applied?

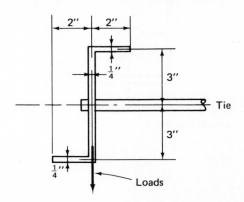

8.9. If the lateral ties are removed (refer to Problem 8.8), what is the maximum stress in the zee bar for the same loads that were applied in Problem 8.8?

8.10. Using a steel with $F_y = 50$ ksi, check the adequacy of the laterally unsupported zee section (refer to Problem 8.9) by use of the adaptation of the AISC interaction formula used in Example 8.8, utilizing the equivalent column allowable stress procedure for bending about the minor principal axis.

8.11. For a channel section with the same I_x and same area as the zee section of Problem 8.8, and with the same loads and span, ties must now be provided to prevent twisting, as shown, again at the third points where the vertical loads are applied. For what force must the ties be designed and in what direction are the tie forces applied?

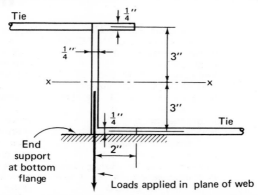

8.12. Design a welded channel girder of the type shown in Fig. 8.10 for a span of 30 ft to support third-point 80-kip loads 2 in. from the centerline of the web with the channel proportioned so that the loads pass through the shear center to eliminate torsion. E70 electrodes and A36 steel may be assumed.

8.13. Design members supported at four points as shown by four different procedures. Assume full lateral support and use W sections of A36 steel. Option (c) requires analysis of a continuous beam by moment distribution or equivalent procedure and should be omitted if necessary. Design is for all spans fully loaded with a live load of 3 kips/ft.
(a) Three separate simply supported spans.
(b) End spans cantilevered over interior support as shown.
(c) A single continuous beam of uniform cross section utilizing criteria of AISCS, Sec. 1.5.1.4.1.
(d) Same as (c), but by plastic design.
Compare the weights of the four alternative designs and discuss the merits of each.

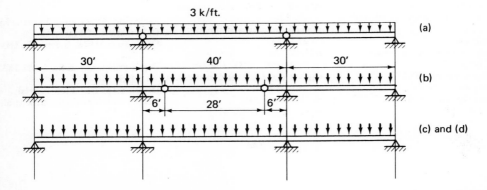

INDEX

A

AISCM, 2
AISCS, 2
Allowable beam load tables, 59
Allowable stress design, 8, 112, 235
Allowable stresses:
  bending, 47-53, 59, 191
  compression, 95
  high-strength bolts, 132, 134
  pins, 25, 138
  reduced stress in flange, 185-186
  rivets, 131, 133
  shear, 53-54
  stress range, 26
  tensile, 25
  unfinished bolts, 131-133
  welds, 142-145
Amplification factor, 114
Angle connections, 146-149
Angle struts, 90-91, 104
Axial compression, equivalent, 113, 115-116

B

Bars, 20
Base plates, 95
Basic column strength, 83-86
Battened columns, 94
Beam-columns (see Columns under combined stress)
Beam equation, 42
Beams, 36-37, 209-240
Beams, allowable load tables, 59
Beam-to-column framing, 39, 128-129, 157-167
Bearing plates, 58
Bearing in rivets or unfinished bolts, 132
Bearing stiffeners, 193-195
Bearing supports, 57
Bearing type connections, 131-132, 159-160

Bending stress, 42, 213-223
Biaxial bending, 56, 73, 223-232
Bolted connections, 129-138
Bolts, 130-132
Box sections, 49, 90, 212, 222
Bracing (see Lateral supports)
Bracket connection, 136-138, 148, 154-156
Brittle fracture, 5
Buckling, beam (see Lateral-torsional buckling)
Buckling, column, 83-84
Built-up beams (see Plate girders)
Built-up members, 24, 92-94 (see also Plate girders)
Butt splice, 135-136

C

Cables, 19
Cantilever beams, 39
Channel beam, 232-235
Column base plates, 95
Column buckling, 83-86
Column Research Council, 85
Columns:
  under axial load, 82-108
  under combined stress, 2, 111-127
  with lacing, battens, or perforated cover plates, 92-93
Column splices, 95
Column strength, 83-86
Column web stiffeners, 170
Combined bending and tension, 18, 126
Combined bending and torsion, 213-223
Combined shear and tension in bolts and rivets, 133-134
Combined shear and tension stress, 149, 219
Common bolts, 131
Compact sections, 45, 47

Compact sections, transition to noncompact, 48
Composite design, 235
Compression elements, 47-48
Compression members (see Columns)
Connections:
  beam-to-column, 171-174
  bolted, 129-138
  concluding remarks, 174
  eccentrically loaded, 150-156
  end plate, 162-163
  flange-to-web, 195-196, 204
  flexible, 128
  girder elements, 195-197, 204-206
  moment-resisting, 167-174
  moment and shear, 168
  pinned, 138-140
  rigid frame, 128
  riveted, 129-138
  seated beam, 162-167
  semi-rigid, 129
  shear, for frames, 157-167
  stiffeners, 196-197, 205-206
  welded, 140-150
Construction methods, 14
Continuous beams, 235-240
Cracks, 5, 26
Crane runway beams, 74-79
Curvature, 41, 42, 44
Cut-off cover plates, 186-188

D

Deflection limitations, beams, 55
Deformation of beam, 42
Depth-span ratios, beam, 55, 180-181
Depth-thickness ratio, 181-183, 185
Design steps (see Flow charts)
Details and drawings, structural, 12, 57

Developments of structural design, 6
Dishing failure, 23
Double angle struts (*see* Angle struts)
Ductility, 4

E

Eccentric loads, 19, 150-156
Economy of structural design, 9
Effective fillet weld throat dimension, 143
Effective length of columns, 84, 86-88
Elastic bending, 41
Elastic design (*see* Allowable stress design)
Elasticity, 4
Elastic limit, 4
Endurance limit, 4
Equivalent axial compression loads, 113, 115-116
Euler formula, 83, 85, 114
Examples (*see* Illustrative examples)
Eyebars, 22, 30-31

F

Fabrication methods, 13
Factor of safety, 8
Factor of safety, column, 95-96
Failure:
  dishing, 23
  fatigue, 26
  net section, 23
  pin-connected plate, 23
Fatigue, 4, 26
Fillet welds, 141-147
Flange moment due to twist, 215-216
Flange selection, 183-189
Flow charts:
  allowable bending stress, 59-62
  allowable compression stress, 97
  column selection for axial load, 98-99
  column selection for combined axial compression and bending loads, 117-118
  combined axial tension and bending loads, 119-120
  intermediate stiffeners selection, 192
  plate girder flange selection, 189
  tension member selection, 27, 119-120
Fracture, 5, 23

Framing angle connections, 157
Friction-type connections, 132-134, 160

G

Girders (*see* Plate girders)
Groove welds, 141-143
Gusset plates, 128

H

Hanger loads, 234
High-strength bolts, 132 (*see also* Friction-type connections)
History of structural design (*see* Structural design development)
Hollow cylinder in pure torsion, 210
HP shapes, 91

I

Illustrative examples:
  beams, 65-79
  columns under axial load, 99-109
  columns under combined stress, 121-126
  connections:
    bracket connection, 136-138, 154-156
    butt splice, 135-136, 148
    framed beam shear connections, 159-162
    lap joints, 134-135, 147
    pin connection, 139-140
    seated beam, 162-167
    semirigid beam to column, 171-174
  plate girders, 197-206
  tension members, 28-34
  torsion, 217-218
Incomplete lateral support, 51
Inelastic behavior of beams, 44-45
Initial tension in high-strength bolts, 132
Interaction formulas for column design, 113-116
Intermediate stiffeners, 189-193

L

Laced column, 92-93
Lacing bars, 92-93
Lap joints, 134-135, 147
Lateral support force, 228
Lateral support requirements, beams, 54
Lateral supports (restraint), 40, 48, 50-51, 53-54, 68, 93, 198, 227, 232

Lateral-torsional buckling, 51, 223
Layout, structural, 12
Load and support details, 57
Load-bearing members, 10
Load-factor, 8, 46
Loading condition, 26
Loading cycles, 26
Local buckling, 83, 193
Local crippling in web, 193
Local stress, 26

M

Maintenance requirements, 12
Maximum moment (*see* Plastic moment)
Members:
  beam-columns, 111-126
  beams, 36-79
  built-up, 24
  columns, 82-108
  plate girders, 178-206
  tension, 17-34
Miscellaneous sections, 51
Modular construction, 12
Modulus of elasticity, 4
Moment diagrams, 38
Moment of inertia, 224-227
Moment-resisting connections, 167-174
Moment and shear connections, 168
Moving loads, 75

N

Net section, 23-24
Noncompact members, 48

P

Parallel-axis theorem, 224
Perforated cover plates, 92-93
Permissible stress (*see* Allowable stresses)
Pin connections, 22, 138-140
Pipes, steel, 89-90
Planning and site exploration, 11
Plastic design, 41, 44, 46, 235-240
Plastic modulus, 46
Plastic moment, 44
Plate girders, 75, 178-180
  bearing stiffeners, 193-195, 203-204
  flange plate selection, 183-189, 199-201
  flange thickness change, 186-187
  flange-to-web connection, 195
  intermediate stiffeners, 189-193, 201-203

Plate girders *(cont.):*
  reduced allowable tensile stress, 185, 191
  span-depth ratio, 180
  splices, 196
  stiffener connection, 196
  tension field action, 181-182
  types of, 178-180
  typical girder flanges, 186
  web plate selection, 180-183, 198-199
Plates, pin-connected, 22
Ponding, 55
Principal axes, 224-227
Product of inertia, 224-227
Prying action in connection, 169-170

R

Rectangular or round section, solid, 49
Repeated load:
  beam under, 56, 71, 188
  design for, 26, 34, 188
Residual stress, 85
Rigid connections *(see* Moment-resisting connections)
Riveted connections, 129-138
Rivets, 131
Rods, 20
Roof deflection, 55
Round bars, 89

S

Safety factor, column, 95-96
Safety of structures, 10
Seat angle connections, 156-159
Secondary members, 10
Section modulus, 43
Semi-rigid connections *(see* Moment-resisting connections)
Service and maintenance requirements, 15
Shape factor, 45-46
Shapes, 92
Shear center, 232-235
Shear connections for frames, 157-167
Shear and moment diagrams, 38
Shear in rivets or unfinished bolts, 132
Shear stress, 38, 43
Simple beams, 36-37, 66

Simple bending theory, 40
Simple framing, 128
Single angle struts *(see* Angle struts)
Site exploration, 11
Slenderness ratio, 25, 84-85
Solid sections, 49, 89, 99-100
Span-depth ratios, beam, 55
Special topics in beam design, 209-240
Specifications, 1-6, 8
Steel columns, types of, 88-94
Steel pipes, 89-90, 101-102, 210
Steel, structural, 4
Steps *(see* Flow charts)
Stiffened plate elements, 47-48
Stiffeners:
  bearing, 193-195
  column web, 170
  connections, 196-197
  intermediate, 189-193
Strain, 5, 41
Stress category, 26
Stress concentrations, 26
Stress distribution after buckling, of web, 185
Stresses:
  bending, 41-42
  on fillet weld throat, 144
  shear, 43, 212
  tensile, 25
Stress range, 26
Stress transmittance from flange to web, 195-196
Strong plane, 40
Structural design:
  allowable stress method, 8
  economy, 9
  historical development, 6
  load-factor, 8
Structural intuition, 3
Structural layout, details, and drawings, 12
Structural safety, 10
Structural shapes, 6, 24
Structural steel, 4
Structural tees, 91, 105
Structural tubes, 90, 102-103, 222
Structure and its parts, 2, 3
Submerged arc welds, 145
Support conditions of girder, 193
Support details, 57
Suspension bridges, 22
Systems engineering, 2

T

Tees, structural, 91, 105
Tensile stress, 5, 25
Tensile stress area, 132
Tension field action, 181-182
Tension members, 17-28
Torsion, 51, 52, 209-223
Torsion constants, 212-213
Toughness, 4
Tubes, structural, 90, 210
Tubular members *(see* Steel pipes *and also* Structural tubes)
Twist angle, 215-216

U

Ultimate plastic load *(see* Plastic design)
Ultimate tensile strength, 5
Unfinished bolts, 131 *(see also* Bearing type connections)
Unstiffened plate elements, 48
Upset rods, 20

V

Vertical web compression, 182

W

W shapes, 38, 91-92, 106-108, 121-126, 214, 217-222
Warping *(see* Torsion)
Weak plane, 40
Web buckling, 193
Web compression, vertical, 182
Web depth (girder), 180
Web plate, selection of, 180-183
Web splices, 196-197
Web thickness (girder), 181
Welded connections, 140-150
Welded joints, types of, 141-142
Welded shapes, 91
Welding process, 140-141
Weld nomenclature, 143
Weld size, 145-146
Wide-flange shapes *(see* W shapes)
Width-thickness ratios, 47-48, 94-95, 184-185
Wire rope, 19

Y

Yield moment, 44
Yield point, 4